AF454351

THE AUTHORS

Dr. K.P. Biswas, M.Sc., Ph.D., D.F.Sc. (Bombay), E.F. (West Germany),F.Z.S., F.A.B.S. (Kolkata) initiated the work on Electro-physiology and Electric Fishing, first in India, in 1961 at Odisha. He designed and developed electric seine, electric fish catcher, electric gill net, electric dip net and a new technique of harmless electric fishing, for which patent applications have been filed. He guided eleven M.Sc. students and one Ph.D. student of Calcutta University and four M.F.Sc. students of West Bengal University of Fishery and Animal Sciences for their thesis work on electrophysiology and fishing with electricity. His latest book is "Five inventions in electrofishing in India",published on 2021.

Miss Namrata Basu, M.Sc. from Calcutta University and a research scholar of Electrophysiological Laboratory, is actively associated with the studies on the application of galvanic current on different plant species, besides electrofisheries work for more than four years and took keen interest in preparing the manuscript of the book.

Miss Arpita Roy, M.Sc. from Calcutta University and a research scholar of Electrophysiological Laboratory, is actively associated with the studies on the application of galvanic current on the germination and plant growth for more than two years and took interest in preparing the manuscript of the book.

Application of Galvanic Current in Germination and Plant Growth

— *Authors* —

K.P. Biswas

Namrata Basu

Arpita Roy

2022

Daya Publishing House®
A Division of

Astral International Pvt. Ltd.
New Delhi – 110 002

© 2022 AUTHORS
ISBN: 9789354613852

Published by   :   **Daya Publishing House®**
A Division of
Astral International Pvt. Ltd.
– ISO 9001:2015 Certified Company –
4736/23, Ansari Road, Darya Ganj,
New Delhi-110 002
Ph. 011-43549197, 23278134
E-mail: info@astralint.com
Website: www.astralint.com

The book is dedicated to the memory of
Late Dr. Kamakhya Pada Biswas and
Late Mrs. Manju Biswas
for their inspiration as long as they were
alive.

...♋ Remembrance ♌...

Inspiration, great sense of humour and a true guide, these are the words which really describes Dr. K.P. Biswas. He had a large stock of stories to tell, all that he experienced throughout his life. Experiencing a storm full night in open ocean in a trawler, being alone in Nicobar sea beach for months, getting an unique opportunity to observe coconut crab in close proximity, etc. these are some of the stories we got thrilled and amazed to hear from him. We were his students, listeners and sometimes his friends. We knew him not only as a teacher, researcher but also as a great human being, a loving husband, a caring elder and a really good cook.

His contribution in the field of electro-fishing in India is remarkable. Throughout his working life he wrote more than twenty books about fisheries and published more than hundred papers in nationally and internationally renowned journals. He loved to explore and study new topic from different fields. At the age of 85, he was still not ready to retire from his research work and continued to invent many new techniques for the advancement in the sectors of fishery as well as agriculture.

Every time he came up with new ideas that really inspired us. He always taught us that negative result is also a result, so don't get upset and try until you have the answer. Under his influences, we also learnt to maintain punctuality, handle adverse situation, and take responsibility. He was our true guide in academy as well as in lession of life. We feel ourselves very lucky for getting the opportunity to have him in our life, though for a very limited time.

In his last two years he was mainly working on galvanic farming in which he was observing the effects of galvanic current on the germination and growth of plant. This book is based on the basic findings from this research. This work needs more trials and thorough study for a longer time. But meanwhile he left us and his work remain unfinished which need to be carried out. Lastly we can only say, we really miss you sir and we will try to follow the path that you have shown us.

His students,
(Namrata Basu, Arpita Roy & Amitavo Roy)
& Co-worker, Prof. N. A. Talwar

Acknowledgement

Authors gratefully acknowledge Prof. N. A. Talwar for his encouragement and interest in this new research that gave us immense boost throughout our experimental works and, Mr. Amitavo Roy, who was actively associated and carried out many experiments at the initial stage and, Mr. Mayur Goswami & Mr. Prayas Auddy, for their help in preparing the manuscript of the book.

The authors deeply grateful to their families for their moral support and, would like to express their gratitude to Astral Publication Ltd. for their aid and co-operation in publishing this book.

Preface

Higher percentage of germination of Bengal gram (*Cicer arietinum*) seeds by application of direct current (DC) stimulation and faster growth with higher rates of branching was reported by Biswas (2004).

The electrical stimulation was given in a low voltage direct current, obtained either from a battery pack or through a step down transformer and rectifier from the service line. The input voltage to the soil and water around germinating seeds and growing seedlings was restricted to a range of 6 to 24 volts. The results of those experiments were found to be promising with regards to higher rate of germination and profuse branching giving a bushy shape to the treated plants.

While experimenting with the effect of galvanic current on blood worm (*Tubifex* sp.), it was noticed that quite a substantial quantity of galvanic current can be generated by different metallic electrode combinations. Tests were carried out with different metallic electrodes to create homogeneous galvanic field in water, moist cotton and soil to obtain maximum potential difference (mV) and intensity of current (μA) between the electrodes. The highest potential difference and intensity thus obtained with suitable electrode combinations and suitable conducting media was to the tune of 1volt (1000mV) and a little more than 1mA (1000 μA).

Since the galvanic current produced through any media is very small (although measurable with sensitive meters). There was doubts of their characteristics on applying load. Accordingly a milli-voltmeter with internal resistance of 1.044 kilo ohms and the capacity to measure 0 - 400 mV was applied, first for a short period and thereafter, round the clock. It was observed that the galvanic current generating capacity of the electrodes gradually declined with the passage of time and came to zero with respect to mV. But on further continuing the load, the potential difference (mV) between the electrodes slowly increased, not to the full scale, but with changing polarity.

Combinations of copper and aluminium, copper and zinc, zinc and aluminium and copper and galvanized iron as electrodes were used, as these combinations gave comparatively higher current intensity and potential difference than the electrodes of the same metal.

K.P. Biswas

Namrata Basu

Arpita Roy

Contents

Introduction

An apparatus that is used to generate electricity from a spontaneous redox reaction or conversely that uses electricity to drive a non spontaneous redox reaction is called an electro - chemical cell.

There are two type of electro - chemical cells; galvanic cells and electrolytic cells. Galvnic cells are named for the Italian physicist and physician Lugi Galvani (1737 - 1798), who observed that dissected frog leg muscles twitched when a small electric shock was applied, demonstrating the electrical nature of a nerve impulses. A galvanic (voltaic) cell uses the energy released during a spontaneous redox reaction ($\Delta G<0$) to generate electricity. This type of electro chemical cell is often called a voltaic cell after its inventor, the Italian physicist, Alessandro Volta (1745 - 1827). In contrast, an electrolytic cell consumes electrical energy from an external sources, using it to cause a non spontaneous redox reaction to occur ($\Delta G<0$). Both types contain two electrodes, which are solid metals connected to a external circuit that provides an electrical connection between the two parts of the system. The oxidation half-reaction occur at one electrode (the anode), and the reduction half-reaction occur at one electrode (the cathode). When the circuit is closed, electrons flow from the anode to the cathode. The electrodes are also connected by an electrolyte, anionic substance or solution that allows ions to transfer between the electrode compartments, thereby maintaining the system's electrical neutrality.

Energy released by spontaneous redox reaction is converted to electrical energy

Oxidation half-reaction;

$$y \rightarrow y^+ + e$$

Reduction half reaction;

$$Z + e \rightarrow z$$

Overall cell reaction;

$$y + Z \rightarrow y^+ + z$$

Empirical studies have been made since 31/05/2016 on the application of galvanic current around germinating seeds and growing seedlings to know if the induced galvanic current, generated from a spontaneous redox reaction, with the aid of electrodes of similar and dissimilar metals have any effect on the germinating plant seeds and growing seedlings and plants.

Experiments have been designed to determine the effect of galvanic current in the three ways, germination of seeds, where not only the percentage of germination was observed, but also the germination period, that is, time taken to germinate the first seed and the maximum number of seeds. Potable water, garden soil and sterilized moist cotton in air tight glass were used as media of galvanic current production.

A combination of copper and aluminium plates, copper and zinc plates, zinc and aluminium plates and copper and galvanized iron plates as electrodes were used, since these electrode combinations gave comparatively higher current and potential difference in the field.

The production of galvanic current (mV and μA) were recorded after every 12 hours initially and thereafter every 24 hours subsequently.

Seeds of maize, gram, bean, tomato, cucumber, paddy were used for these tests. A problem was faced on using water and garden soil as germinating media. Maggot infestation on soaked seeds in water media and insect infestation in soil media both in experimental and control greatly reduced the germination results. Moist sterilized cotton media,covered under a glass jar gave satisfactory results with respect to percentage of germination.

Galvanic current intensity was changed by shortening and lengthening the distance between the electrodes in the conducting media. Accordingly, three sets of tests were made with full length, half length and quarter length of the trough to vary distances between the generating electrodes.

The second stage of experiments were to shift the seedlings either to an intermediate bed of soil or glass tank, wooden troughs or earthen pots to grow either with a galvanic field around them or without galvanic field. Records were kept on the growth of seedlings in this new environment by measuring the length of shoots, number of leaves and thickness of stems.

In the third stage, the seedlings from intermediate beds were transplanted to grow in ground soil with or without galvanic field coverage. In this final stage, the growth of plants were recorded at intervals by measuring the length of shoots, thickness of the stems, branching frequency, flowering and fruiting.

The growth of treated and untreated plants with respect to length, thickness of stem, foliage and bearing fruits were observed till the fruits were ripe and matured in some cases. The fruit bearing capacity, the size and weight of fruits when ripened or matured and the weight of total biomass of both treated and untreated plants were compared with each other to know best of the process.

The promising results of galvanic current intervention during germination of seeds and plant's growth advocate for a possibility of adapting this technique in future plant production. As the technique is a new one, the inventor wishes to name the technique as "M and K Technique of galvanic current intervention to plant production".

Characteristics of Galvanic Current

In the simplest way, galvanic current is a low level, continuous current that flows in one direction known as direct current (DC), just like the electricity obtained from a battery. This differs from high frequency that uses an alternating radio current, called the "Tesla current".

Galvanic current is constant and direct current, having a positive and negative pole and produce chemical and ionic reactions.

The wave of galvanic devices is monophasic, meaning it has only one peak that is repeated over times, galvanic stimulation uses direct current to create unidirectional continuous current. Galvanic stimulations are used to produce ionic movement within tissues toward one of the electrode.

Interrupted current is a direct unidirectional current. A regular pause is given between the stimulations in interrupted galvanic current (1GC).

Galvanic current contains negative ions and treatment gel (containing anti-aging and other ingredients) that is applied to the skin contains positive ions. The current pushes the products deeper into the skin. The product work hard in neutralizing the toxins and harmful free radicals of the skin.

The process of forcing positive substances into deeper tissues using galvanic current from the positive towards the negative pole tightens and calms the skin. In constant direct current, it uses a positive and a negative pole to produce chemical reactions (desincrustation) and ionic reactions (iontophoresis).

Galvanic shock occurs when the filling of one tooth containing amalgam, touches the filling of another tooth, containing another metal. Saliva facilitates the shock by functioning as an electrolyte.

Faradic current differs from galvanic current in that, a faradic type current is a short duration interrupted direct current. They have a pulse duration of 0.1 to 1 millisecond (ms) and a frequency of 50 to 100 Hz. But in galvanic type of current, the duration of interruption can be adjusted, a duration of 100ms is commonly used, with a frequency of 30 per minute.

Micro current is an alternating current. Unlike galvanic current both the probes in true micro current are placed on the face simultaneously and each probe provides both the negative and positive charges. Micro current is different from galvanic current, since galvanic current works with the chemical reaction of the solution and the current.

As early as 1804, the physicist Giovanni Aldini applied galvanic (direct) current to treat individuals suffering from melancholia.

Galvanic current is used primarily for facial muscles. A galvanic current uses mild electric current that consists of positive and negative ions to stimulate local muscles. The outcome of a galvanic current is maintaining the muscle physiology.

Applications and Effects of Galvanic and Direct Current

Galvanic or direct current is a form of electrotherapy treatment. The electrical current, restricted to a safe, low-voltage level is applied to the body via electrodes placed on the skin. Galvanic treatment use gentle, safe electrical currents to intensify the effects of profesional skin care products and give skin a thorough cleanse.

As early as 1804 physicist, Giovanni Aldini applied galvanic (direct) current to treat individuals suffering from melancholia.

Electrical stimulation promote tissue healing. Medical galvanism, or the use of galvanic stimulation, uses direct current modalities that deliver a unidirectional, uninterrupted current flow within the tolerance of the patient and without destruction of the tissue. This type of modality can be used to directly stimulate muscles following a nerve injury, to produce ionic change within the tissues and decrease edema or to introduce typically applied medications into the skin (iontophoresis). The purpose of this electrical stimulation is primarily for the vasomotor effects; *i.e.* increased circulation. Under the electrodes, ions accumulate in the skin. The sensation experienced acts as a physiological stimulus to the sensory nerve endings, producing reflex vasodialation. These vasomotor effects can assist in resolution of inflammation, relief of pain, and reduction of interstitial edema through electro-osmosis and the shifting of water toward the electrical cathode.

Direct, uninterrupted electric current tends to be quite uncomfortable and may cause superficial skin burn. For this reason, a modification of the technique has been developed whereby the current is applied in an alternating or "pulsed" manner, termed as high voltage pulsed galvanic current (HVPGS). Although the main use of HVPGS in sports rehabilitation is for relief of pain, it can also be used to aid tissue healing.

1. Methods of Neurotherapy

From basic neurophysiology, it is known that a constant direct current (DC) shifts membrane potential of neurons towards either hypo or hyper polarization depending on the direction of current. Although the effect of electrical current on living tissues has been known of centuries, only in the middle of the 20th century systematic research of this phenomenon has been initiated. The evidence that a weak scalp DC can induce prolonged changes in brain excitability, opened new approaches to the management of neurological conditions. However even now-a-days, whereas ample experimental data are available on brain polarization in animals, little is known about how weak DC applied through the scalp affects brain excitability in humans. Now-a-days, the procedure is called tDCS.

Weak direct current in tDCS change membrane potentials of cortical neurons only slightly. The current are much smaller than those used in a so called electro convulsion therapy (ECT).

ECT was discovered in 1930s by Italian scientists Bini and Cerletti. In those days ECT appeared as a fundamental breakthrough in the management of mental disorders. ECT produced the marked and consistent improvement in patients. ECT fundamentally differs from tDCS. Whereas ECT injects strong currents inducing convulsion activity in the brain, tDCS induces much smaller neuronal activity without inducing seizures.

Available evidence indicates that unlike ECT, tDCS causes no memory disturbance or loss of consciousness.

2. In Dentistry

The presence of metallic restorations in the mouth may cause a phenomenon called galvanic action, or galvanism. This results from a difference in potential between dissimilar fillings in opposing or adjacent teeth. This fillings, in conjunction with saliva or bone fluids, such as, electrolytes make up an electric cell. When two opposite fillings contact each other, the cell is short circuited and if the flow of current occurs through the pulp, the patient experience pain and the more anodic restoration may corrode. A single filling plus the saliva and bone fluid may also constitute a cell of liquid junction type.

Studies have indicated that relatively large currents will flow through metallic fillings when they are brought into contact, probably as a result of polarization of the cell. The magnitude of the voltage, however, is not a primary importance because indication support the fact that the sensitivity of the patient to current has a greater influence on whether pain is felt. Although most patients feel pain at a value of 20 and 50 µA, some may feel pain at 10 µA, whereas other do not experience it until, 110 µA is developed.

The galvanic currents developed from the contact of two metallic restorations depend on their composition and surface area. An alloy of stainless steel develops a higher current density than either gold or cobalt-chromium alloys when in contact with amalgam restoration.

The principal method of avoiding corrosion should be apparent, dissimilar metals or alloys should not be allowed to come into contact.

A metallic crown should not have a electrical connection to silver endodontic points in the same tooth. On contact, such a circuit would be closed and the external circuit, passing as it does partially through oral tissues via the pulp in each tooth, may result in a corrosion current, large enough to cause significant shock, detectable at least, by the pulp. This is the basis of what is called "dental galvanism".

An electrical circuit may be formed in the mouth when dissimilar metalic restorations are brought into occlusal contact, resulting in a shock.

If one experience galvanic shock, the symptoms should disappear within 24 hours. If the galvanic shock stems from two unlike metals placed next to each other, such as fillings, one of the filling may need to be replaced with a non - metallic substance, such as porcelain.

3. Clinical Significance of Galvanic Current

Galvanic direct current used to stimulate desired muscles and for ion transfer (iontophoresis). High voltage pulsed galvanic current is used to relieve pain and relax muscle spasm.

Iontophoresis is a process of infusing water-soluble products into the skin with the use of electric current, such as the use of positive and negative poles of a galvanic machine.

Plant Physiology

Plants

An organism that carries out photosynthesis, has cellulose cell walls, complex cells, and is immobile. A few parasitic plants have lost the ability to photosynthesize, but are still considered to be plants.

The groups that are always classified as plants are the bryophytes (mosses and liverworts), pteridophytes (ferns, horse tails and club mosses) gymnosperms (conifers) and angiosperms (flowering plants). The angiosperms are further classified into monocotyledons (orchids, grasses, lilies) and dicotyledons (apple, oak, rose *etc.*).

Plants organs include roots, stems, leaves and reproductive structures. Each plant organ performs a specialized task in the life of the plant. Roots, stems and leaves are all vegetative structures.

Roots are important organs in all vascular plants. Most vascular plants have two type of roots, primary roots that grow downward and secondary roots, that broken out to the side. Together, all the roots of a plant make up a root system.

Plants have a life cycle, just like other animals. The plant life cycle describes the stages the plant goes through from the beginning of its life until the end, when the process starts all over again. The plant's life cycle starts with germination of the seed. All is needed are water, oxygen and the right temperature for the seed to germinate. The major stages of the flowering plant life cycle are the seed germination, growth, reproduction, pollination and seed spreading stages.

There are many things that plants need to grow, such as, water, nutrients, air, light, temperature, space and time. With warmth, oxygen and nutrients from the soil, the seed will pop open and sprout. Once the leaves and roots grow, the plants can go through the process of photosynthesis, The process that plants use to breath and to make food using sun light.

Physiology

Physiology is the study of normal functions within living creatures. It is a solo-division of biology covering a range of topics that include, organs, anatomy, cells, biological compounds and how they all interact to make life possible.

Plant physiology is a sub-discipline of Botany, concerned with the functioning or physiology of plants. Closely related fields include plant morphology, plant ecology, phytochemistry, cell biology, genetics, biophysics, and molecular biology.

Plant physiology is the study of how different parts of plants function. It includes many aspects of plant life, including nutrition, movement and growth. It studies the ways in which plant absorb minerals and water, grow and develop, flower and bear fruits. Research in plant physiology provides a scientific basis for the rational planting of corps with regard to soil and climatic conditions.

Crop physiology in plant is applied to a cropping situation. In that way it can be considered a sub-discipline of plant physiology, but also agronomy, crop physiology is more applied and "macro" oriented than plant physiology. Crop physiology often involves the study of population versus crop.

Fundamental process, such as, photosynthesis, respiration, plant nutrition, plant hormone function, tropism, nastic movements, photoperiodism, photomorphogenesis, circadian rhythm, environmental stress, physiology, seed germination, dormancy and stomata function and transpiration are all physio chemical process.

Nitrogen promotes the growth of foliage, while phosphorus and potassium support the growth of strong roots, stems, flowers and fruits. All these three elements are essential in the photosynthetic process, which uses sun's energy to change carbon di oxide and water into the sugars and starches that feed plants.

There are three type of plant tissues, namely, dermal, ground and vascular tissues. Dermal tissue covers the outside of a plant in a single layer of cells, called epidermis. Epidermis helps the exchange of matter between the plant and environment. Parenchyma, collenchyma and selerenchyma are the examples of ground tissue. Parenchyma is mainly found in the soft parts of leaves and roots for the purpose of photosynthesis and food storage respectively. Collenchyma and selerenchyma, both are involved in shoot support but the former is found in areas of active growth, while the latter in areas where growth has ceased. Vascular tissue consists of xylem and phloem. Xylem transport water, nutrients from roots to leaves and phloem send food from leaves to other parts of the plant.

The three phases of cell growth are cell division, cell enlargement, and cell differentiation. The first two stages increase the size of the plant cell, while, third stage brings maturity to the cells. Differentiation is a process during which cells undergoes structural changes in the cell wall and protoplasm.

Plant Cell Biology

A plant cell is an eukaryotic cell that contains a true nucleus and certain organelles to perform specific functions. However, some of the organelles present in plant cells are different from other eukaryotic cells (animals).

The organelles found only in plant cells include, chloroplast, cell wall, plastids and large contractile vacuole. The chloroplast contains a green pigment chlorophyll that is responsible for the process of photosynthesis.

Plant cells are basic building blocks of plant life, and they carry out all the functions necessary for their survival. Photosynthesis, the making of food from light energy, carbon dioxide and water occurs in the chloroplasts of the cell. In ideal growth conditions, the bacterial cell cycle is repeated every 30 minutes. Only a few types of eukaryotic cells can grow and divide as quickly as bacteria. Most of the growing plant and animal cells take 10 to 20 hours to double in numbers and some duplicate at a much slower rate.

Parts of Plant Cell

- ☆ Cell wall - It is a rigid layer, which is composed of cellulose, glycoproteins, lignin, pectin, and hemicellulose.
- ☆ Cell membrane - It is a semi-permeable membrane that is present within the cell wall. It control what goes into and out of the cells. It is also known as plasma membrane and is present adjacent to the cell wall.
- ☆ Nucleus - Controls all activities, involves with reproduction and protein synthesis, and contains cell's in genetic formation (DNA).
- ☆ Mitochondria - Breaks down food to release energy for the cell.
- ☆ Endoplasmic reticulum -Transport materials within the cell and process lipids.
- ☆ Nuclear membrane - A nuclear membrane is a double membrane that encloses the cell nucleus. It serves to separate the chromosomes from the rest of the cell. The nuclear membrane includes an array of small holes or pores to permit the passage of certain materials, such as, nucleic acids, and proteins between the nucleus and cytoplasm.
- ☆ Cytoplasm - A thick solution that fills each cell and is enclosed by the cell membrane. It is mainly composed of water, salts and proteins. In eukaryotic cells, the cytoplasm includes all of the material inside the cell.
- ☆ Nucleoplasm - A type of protoplasm that made up mostly of water, a mixture of various molecules and dissolved ions. It is completely enclosed within the nuclear membrane or nuclear envelop. It is a highly gelatinous, sticky liquid that supports the chromosomes and nucleoli.

☆ Nucleolus - The largest structure in the nucleus of eukaryotic cells. It is best known as the site of ribosome biogenesis. Nucleoli also participate in the formation of signal recognition particles and play a role in the cell's response to stress.

☆ Ribosome - A minute particle consisting of RNA, and associated proteins found in large numbers in the cytoplasm of the living cells. They bind messenger RNA and transfer RNA to synthesize polypeptides and proteins.

☆ Lysosome - A lysosome is a membrane - bound cell organelle that contains digestive enzymes. Lysosomes are involved with various cell processes. They break down excess or worn out cell parts. They may be used to destroy invading viruses and bacteria.

☆ Peroxisomes - It is a membrane - bound organelle found in the cytoplasm of virtually all eukaryotic cells. Peroxisomes are oxidative organelles. Here, molecular oxygen serves as a co-substrate, which helps in the formation of H_2O_2 frequently.

☆ Chloroplasts - Chloroplast are organelles that conduct photosynthesis, where the photosynthetic pigment chlorophyll captures the energy from sunlight, converts it, and stores it in the energy-storage molecules ATP and NADPH while freeing oxygen from water in plant and algal cells.

☆ Golgi apparatus - A complex of vesicles and folded membranes within the cytoplasm of most eukaryotic cells, involved in secretion and intracellular transport.

☆ Plastids - Any of a class off small organelles in the cytoplasm of plant cells containing pigment or food.

☆ Central vacuole - The central vacuole is a large vacuole found inside the plant cell. A vacuole is a sphere field with fluid and molecules that stores water and maintains turgor pressure in a plant cell.

Cell components like cell membrane, nucleus, mitochondria, golgi and ribosomes are common in both plant and animal cells. However, it is the presence of three additional parts, namely, cell wall, chloroplast and vacuole which make it a plant cell.

Plant cells which are found in leaves carry out photosynthesis and cellular respiration, along with other metabolic processes. They also stores substances like starches and protein and have a role in giant wound repair.

Plant Cell Types

A typical plant cell consist of a relatively rigid cell wall lined with a cell membrane. Within the cell membrane lie the nucleus and other structures suspended in a liquid matrix called the cytoplasm.

A few plant cells are involved in the transportation of nutrients and water; while others for storing food. The specialized plant cells include parenchyma cells, sclerenchyma cells, collenchyma cells, xylem and phloem cells.

Each part of the cell has a specialized function. These structures are called organelles. Specialized structures in plant cells include chloroplasts, a large vacuole and the cell wall. Chloroplasts are organelles that are crucial for carrying out photosynthesis in the plant cell, using sunlight.

Function of Organelles

The largest organelle in plant cell is the central vacuole containing acid and several types of digestive enzymes. It stores water, waste products, food and other cellular materials.

The cell wall is a covering that protects plant cells and gives them shape. Mitochondria releases energy stored in food. Chloroplast makes sugar, using sun's energy. Although ribosomes are not enclosed within a membrane, they are still commonly referred as organelles in a eukaryotic cell and are considered to be the cells protein factory. Lysosomoes digest unwanted organelles in a process termed "autophagy". The multipurpose lysosome also processes proteins, bacteria and other food the cell engulfed.

Nucleus is a membrane bound organelle containing DNA, the control centre of the cell. Cytoplasm is found inside the cell membrane. It is the gel like substance that surrounds the nuclear membrane.

Both animal and plant cells have mitochondria, but only plant cells have chloroplasts. The process of photosynthesis takes place in the chloroplast. Once the sugar is made, it is broken down by the mitochondria to make energy for the cell.

The endoplasmic reticulum (ER) is an organelle responsible for making both membranes and their proteins. It also aids in molecular transport through its own membrane.

The endoplasmic reticulum is responsible for protein translocation, which is the movement of protein through the cell.

Ribosome is the smallest cell organelle. They are either spherical or granular particles and are scattered freely in the cytoplasm. They are also attached to the surface of endoplasmic reticulum, which are called rough endoplasmic reticulum (RER).

Transport Across the Cell Membrane

Active transport requires chemical energy, because it is the movement of biochemicals from areas of lower concentration to areas of high concentration. On the other hand, passive transport moves biochemicals from areas of high concentration to areas of low concentration; and as such, does not require energy.

Passive transport is the movement of molecules across the cell membrane and does not require energy. It is dependent on the permeability of the cell membrane. There are three main kinds of passive transport, diffusion, osmosis, facilitated diffusion.

> ☆ **Simple diffusion** - Movement of small or lipophilic molecules (O_2, CO_2 *etc.*) from a region of higher concentration to the region of lower concentration.

> ☆ **Osmosis** - Movement of water molecules (dependent on solute concentration).

> ☆ **Facilitated diffusion** - Movement of large or charged molecules via membrane protein (ions, sucrose *etc.*).

When molecules move in this way, they are said to move down their concentration gradient.

Passive transport does not require ATP (energy), but active transport does require energy.

Osmosis is a form of passive transport, similar to the diffusion and involves a solvent moving through a selectively permeable or semi-permeable membrane from one area of higher concentration to an area of lower concentration.

The primary cause of diffusion is random movement of atoms and molecules in a substance.

Movement of Cell Sap

Xylem sap carries soil nutrients (dissolved minerals) from the root system to the leaves, the water is then lost through transpiration.

Sap in plants consists of liquid inside the large central vacuole of a plant cell that serves as storage of materials and provides mechanical support, specially in non-woody plants. It has also a vital role in plant cell osmosis, the cytosol, which is the watery fluid component of cytoplasm.

The cohesive properties of water (hydrogen bonding between adjacent water molecule) allow the column of water to be pulled up through the plant as water molecules are evaporating at the leaf surface. This process has been termed the "Cohesion theory of sap ascentric plants"

Water is passively transported into the roots and then into the xylem. The forces of cohesion and adhesion cause the water molecules to form a column in the xylem. Water moves from the xylem into the mesophyll cells, evaporates from their surfaces and leaves the plant by diffusion through the stomata.

Plant absorbs water from the soil by osmosis. They absorb minerals ions by active transport, against the concentration gradient. Root hair cells are adopted for facing up water and mineral ions by having a large surface area to increase the rate of absorption. A root hair, or absorbent hair, the rhizoid of a vascular plant, is a tubular outgrowth of a trichoblast, a hair-forming cell on the epidermis of a plant root. As they are lateral extensions of a single cell and only rarely branched, they are visible to the naked eye and light microscope.

Calcium is an important constituent of middle lamallae in plants, where it is present as calcium pectate. Hence actively growing regions of calcium like stem tip and root hairs essentially require calcium.

Ion Exchange in Plant Root

Cations ion exchange occurs when nutrient cations are attracted to charged surface of cells within the root, called cortex cells. When cation exchange occurs, the plant root releases hydrogen ion. This cation exchange in the root causes the pH of the immediately surrounding soil to decrease.

Cation exchange capacity is defined as a soil's total quantity of negative surface charges. It is measured commonly in commercial soil testing laboratories, by summing cations (positively charged ions that are attracted to the negative surface charges in soil).

Plant roots show the "suspension effect" and interact with neutral salts with the development of exchange acidity and this determine the anion exchange capacity of the plant roots, as the exchange properties of colloids of root system.

Acid-base properties and the swelling capacity of wheat, lupin and pea root cell walls were investigated (Yermakov,). Roots of seedlings and green plants of different age were analysed by the potentiometric method. The ion exchange capacity (Si) and the swelling coefficient (Kew) of root cell walls were estimated at various pH values (2-12) and the different ionic strength (between 0.3 and 1000 mM). To analyse the polysigmoid titration curves pHi = f (Si), the Gregor's equation was employed. It was shown that Gregor's model fits well with the experimental data. The total number of cation exchange (St cat) and the anion exchange (St an) groups were determined in the root cell walls. The number of the functional group of each type (Sj) was estimated, and the corresponding values of pkaj were calculated. It was shown that for all types of cation exchangeable groups arranged in the cell wall structure, the acid properties are enhanced by the increasing concentration of electrolyte. For each ionogenic group the coefficients of Helfferich's equation [pkaj = f (ckt)] were determined. It was found that the swelling of root cell walls changes with pH, ckt and strongly depends on plant species. Within the experimental pH and ckt range, the swelling coefficient changes as follows lupin>pea>wheat. The obtained results show that for the plant species under investigation the differences in the coefficients originate from;

    (a) The differences in the cross-linking degrees of polymeric chains arranged in the cell wall structure;

    (b) The differences in the number of carboxyl groups; and

    (c) The differences in total number of functional groups.

Based on the estimated swelling coefficients in water; it could be inferred that for wheat the cross-linking degree of the polymeric chains in the root cell walls is higher than those for lupin or pea.

It has been emphasized that the calculated parameters, the equations and the dependencies allow to estimate quantitatively the changes in the ion-exchange capacity of the root cell walls in response to the changes in an ionic composition of an outer solution. The results of these estimation allow to suggest that (a) the root apoplast compartment where accumulation of cations takes place during first stage of cation uptake from an outer medium and (b) the accumulation degree is defined by the pH and ionic composition of an outer solution.

On the basis of literature review and the results of the present experimental study, it was proposed that the changes in the cell wall swelling in response to variances of the environmental or experimental conditions could lead to a change of the water flow through a root apoplast. It has been supported that there is direct relationship between the swelling of the root cell and water flow within the plant root apoplast.

Ion Exchange in Soil

In ion exchange reaction, any of a class of chemical reactions between two substances (each consisting of positively and negatively charged ions) that involves an exchange of one or more ionic components. Ions are atoms or group of atoms, that bear positive or negative electric charge.

Cation exchange in the soil consist of interchanging between a cation in the solution of water around the soil particle and another cation that is stuck to the clay surface. A soil with low cation exchange capacity (CEC) is much less fertile because it can not hold on too many nutrients and they usually contain less in clays.

The effective cation exchange capacity (CEC) is defined as the total amount of exchangeable cation, which are mostly sodium, potassium, calcium and magnesium (collectively termed as bases) in non-acidic soils and bases plus aluminium in acidic soils.

Soil cation exchange capacity is a significant number (CEC) for important soil characteristics. A high CEC value (>25) is a good indicators that a soil has a high clay and/or organic matter content and can hold a lot of cations. A soil with a low CEC value (<5) is a good indicator that the soil is sandy with little or no organic matter and that cannot hold many cations. Loams are the most fertile soil type, made up of roughly equal amounts of sand, silt and clay. Because cations are attached to the negative particles in the soil and that is why cations are difficult to get access to in the soil.

Cation exchange capacity is an inherent soil characteristic and is difficult to alter significantly. It influences the soil's ability to hold on to essential nutrients and provides a buffer against soil acidification. Soils with the higher clay fraction tend to have a higher CEC. Organic matter has a very high CEC.

Piper's method for the determination of anion exchange capacity of soils is modified for use with roots of seedlings, The phosphate absorbed from the ammonium phosphate solution is extracted with alkali and determined colorimetrically. In quadruplicate test, the error is usually 1-3 per cent, but occasionally upto 7 per cent.

Plants Absorbing Cations from the Soil

In the cation exchange process the plant's root absorb many of the nutrients essential for plant growth. The process works through the secretion of exudates by the root hairs which contain positively charged hydrogen ions.

Root hairs form an important surface as they are needed to absorb most of the water and nutrients needed for the plant. They are also directly involved in the formation root nodules in legume plants.

Root hairs are found in nearly all vascular plants, including angiosperms, gymnosperms and lycophytes and they exhibit similar cellular features, suggesting a common evolutionary origin. Root hair cells can survive for 2 to 3 weeks and then die off. At the same time new root hair cells are continually being formed at the tip of the root. This way, the root hair coverage stays same as before. This ensures equal and efficient distribution of the actual hairs on these cells.

Having a large surface area, the active uptake of water and minerals through root hairs is highly efficient.

Ionic Characteristic of the Soil

Most soil particles have a negative charge. The amount of negative charge depends on soil texture,such as sand, silt and clay content, which is directly related to the soil particle surface area. Soils are made up of sand, organic matter, silt and clay particles. Soil with high sand content have low holding capacity for cations compared to clayey and silty soils. Clay and silt particles have negatively charged sites which enable them absorb and hold on to cations.

Anions are atoms or radicals (groups of atoms) that have gained electrons. Since they have now more electrons than protons, anions have a negative charge. For example, chloride ion Cl^-, bromide Br^-, Iodine I- *etc.* An ion is an atom (or group of atoms) with an electrical charge. Anions are one of the two types of ions.

Sulphur is commonly found in the anion sulphate form (SO_4^-), which does not bind to cation exchange sites and is thus mobile in most soils.

Soil pH affects nutrients availability by changing the form of nutrient in the soil. Plants usually grow well at pH values above 5.5. Soil pH 6.5 is usually considered optimum for nutrient availability. Lower pH increases the solubility of aluminium, manganese and iron, which are toxic to plant in excess.

Plant Growth

Factors Affecting Plant Growth

Soil Fertility

Soil fertility refers to the ability of soil to sustain agricultural plant growth, that is, to provide plant habitat and result in sustained and consistent yields of high quality. A fertile soil has the following proportion.

(a) The ability to supply essential plant nutrients and water in adequate amount and proportions for plant growth and reproduction and the absence of toxic substances which may inhibit the plant growth.

(b) Sufficient soil depth for adequate root growth and water retention.

(c) Good internal drainage allowing sufficient aeration for optimal root growth (although, some plants, such as paddy tolerate water logging).

(d) Top soil with sufficient soil organic matter for healthy soil structure and soil moisture retention.

(e) Soil pH in the range of 5.5 to 7.0 (suitable for most plants, but some prefer to tolerate more acid or alkaline conditions).

(f) Adequate concentration of essential plant nutrients in plant-available forms.

(g) Presence of range of microorganisms that support plant growth.

(h) The use of soil conservation practices in order to check soil erosion and other forms of soil degradation, which generally result in a decline in quality with respect to one or more of aspects indicated above.

Nutrient Absorption

The ions regardless of light being present or not, plants absorb nutrients (mainly the metal micronutrients) may serve as components of various enzymes. Other nutrients (mainly nitrogen, phosphorus, magnesium and sulfur along with calcium) become the major components of the plant as proteins, sugars, DNA, chlorophyll and a host of other compounds.

Diffusion is the process by which nutrients spread from areas of high concentration to areas of low concentration. When roots absorb nutrients from soil solution, the concentration of nutrients surrounding the root drops. As a result, nutrients from areas of higher concentration in soil solution migrate towards the root.

Phosphorus is an essential element for all the living organisms, including plants. But it is often the most limiting nutrients, in soil-plant systems. Although most agricultural soils have large amounts of inorganic and organic concentration of phosphorus are available to plants.

Limiting Nutrients

Phosphorus is usually considered as the "limiting nutrient" in plant growth. Limiting nutrient is limiting because not only there is not enough of it, but also there is enough of everything else that an organism needs to allow faster or greater growth excepting the limiting nutrients. Limiting nutrients tend to be one or at best few possible nutrients required by an organism.

Plants roots are able to absorb sugars from the rhizosphere, but also release sugars and other metabolites that are critical for growth and environmental signaling. Reabsorption of released sugar molecules could help reduce the loss of photosynthetically fixed carbon through the roots.

The amount of water in the soil and the climate of the region affect the depth of root systems. Some tree roots probe hundreds of feet deep for searching of water and many trees send roots through cracks in rocks. Roots, are therefore, considered as the smartest part of the plant.

Roles of Roots in Plant Growth

The general response to gravity in plants is well known, roots respond positively, growing down into the soil, and stems respond negatively growing upward to reach the sunlight.

Hydrotropism is a plant growth response to water concentrations. The response can be positive (toward the water) or negative (away from the water). Roots for instance, are positively hydrotropic. For this, to bend and grow the roots, a moisture gradient is essential because plants need water to grow. Similarly the ability to bend and grow the stem towards a photo gradient (light) is essential because plant needs sunlight to prepare food through photosynthesis to grow.

Plants growth response towards gravity is known as gravitropism, whereas the growth response towards light is phototropism. As a result, root cells on upper side of the root grow longer, turning the roots downward into the soil, away from the light. Roots will also change direction when they encounter a dense object such as rock.

Plants grow differently in zero gravity. They will grow away from a light source, regardless of gravitational forces. Waving, however, is significantly different in outer space and the ISS roots curved and waved through their growth medium is a subtler pattern than they would have on earth.

Abundant evidence demonstrates that roots bend in response to gravity, due to a regulated movement of plant hormone auxin, known as polar auxin transport. Auxins are hormone that stimulate growth and are produced in immature parts of plant. They were the first group of hormone studied in plants. Cytokinins are chemicals produced in the roots, which stimulate growth and anti - aging effects.

Plant absorbs water from the soil by osmosis. They absorb mineral ions by active transport against concentration gradient. Root hair cells are adapted for taking up water and nutrient ions by having a large surface area to increase the rate of absorption.

Osmosis uses the difference in concentrations of nutrients between soil and the root to move water and nutrients in plants. Once the water and nutrients are inside the xylem, adhesion and cohesion (capillary action) continue to move the water with nutrients up through the plants.

Chapter 5

Electrochemistry

Electrochemistry is the branch of physical chemistry that studies the relationship between electricity, as a measurable and quantifiable phenomenon and identifiable chemical changes, with either electricity considered as an outcome of a particular chemical changes or vice versa.

Electrochemical process means electrochemical reaction, *i.e.*, any process either caused or accompanied by the passage of an electric current and involving in most cases the transfer of electrons between a solid and a liquid substance.

Electrochemical cells produce a voltage by making the electrons from a spontaneous reduction - oxidation reaction flow through an external circuit. The tendency of the system to go to a lower energy state shows as a voltage (potential energy) difference between the electrodes.

Electrochemistry can afford to control electric field near the plant root as desired. If the electrochemical potential is applied to the root, the signal is supposed to be sensed near the interior of the root, which may possibly stimulate the growth of the plant.

There are two types of electrochemical cells, galvanic also called voltaic and electrolytic. Galvanic cells derive its energy from spontaneous redox reactions, while electrolytic cells involve in non-spontaneous reactions and thus require an external electron source like a DC battery on an AC power source.

Galvanic cells convert chemical potential energy into electrical energy. The energy conversion is achieved by spontaneous ($\Delta G<0$) redox reactions producing a flow of electrons. Galvanic cell, among other cells, is a type of electrochemical cell. It is used to supply electric current by making the transfer of electrons through a redox reaction. A galvanic cell is an exemplary idea of how energy can be harnessed using simple reactions between a few given elements.

Electrochemistry deals with oxidation-reduction reactions that either produce or utilize electrical energy and electrochemical reactions take place in cells. Each cell has two electrodes, the conductor through which electron enter and leave the cell.

Copper electrode is the positive electrode and zinc electrode is the negative electrode. As the cells reactions proceed, atoms of the zinc electrode lose electrons and move into the solution as zinc ions. At the same time, Cu^{2+} ions acquire electrons at the copper electrode.

Since Zinc is placed much above copper in the reactivity series, Zinc is much more reactive than copper in an galvanic cell. Zinc is made the cathode, because it can lose electrons and copper is made the anode as it can gain electrons much easily as compared to zinc due to its low reactivity.

Five important concepts may be borne in mind in this regard;

(a) The electrode's potential determines the analyte's form at the electrode surface;

(b) The concentration of analyte at the electrode's surface may not be the same as its concentration in bulk solution;

(c) In addition to oxidation-reduction reaction, the analyte may participate in other reactions;

(d) Current is a measure of the rate of analyte's oxidation or reduction; and

(e) Cannot simultaneously control current and potential.

Oxidation - Reduction Reaction

For an oxidation - reduction reaction, the potential determines the reaction's position. If an electrode is placed in a solution of Fe^{3+} and Sn^{4+} and adjust the potential to 0.500V, Fe^{3+} is reduced to Fe^{2+}, but Sn^{4+} remain unchanged.

Nernst equation provides a mathematical relationship between the electrode's potential and the concentrations of an analyte's oxidized and reduced forms in the solution. For example, according to the Nernst equation for Fe^{3+} and Fe^{2+} is

$$E = E^0 - RT/nF \log [Fe^{2+}]/[Fe^{3+}]$$

$$= E^0 - 0.05916/1 \log [Fe^{2+}]/[Fe^{3+}];$$

Where E is the electrode's potential and E^0 is the standard - state reduction potential for the reaction $Fe^{3+} = Fe^{2+} + e^-$. Because it is the potential of the electrode that determines the analyte's form at the electrode's surface, the concentration terms in equation mentioned above are those at the electrode's surface, not the concentrations in bulk solution.

The distinction between surface concentrations and bulk concentrations are important. If an electrode is placed in a solution of Fe^{3+} and fix its potential at 1.00 volt, then from the ladder diagram, it can be known that Fe^{3+} is stable at this potential, the concentration of Fe^{3+} remains the same at all distances from the electrode's surface. But if the electrode's potential is changed to +0.500 V, the concentration of Fe^{3+} increases as one move away from the electrode's surface until it equals the concentration of Fe^{3+} in bulk solution. The resulting concentration gradient causes additional Fe^{3+} from the bulk solution to diffuse to the electrode surface.

The solution containing this concentration gradient in Fe^{3+} is called the diffusion layer.

Current is a Measure of Rate of Reduction Reaction

The reduction of Fe^{3+} to Fe^{2+} consumes an electron, which is drawn from the electrode. The oxidation of another material, perhaps the solvent, at a second electrode serves as the source of electron. The flow of electrons between the electrodes serves as the source of electron. The flow of electrons between the electrodes provides a measurable current. Because the reduction of Fe^{3+} to Fe^{2+} consumes one electron, the flow of electrons between the electrodes - in other words, the current is a measure on the rate of reduction reaction.

One important consequence of this observation is that the current is zero when the reaction $Fe^{3+} = Fe^{2+} + e^-$ is at equilibrium.

Electro Chemical Potential around Plant Root

It is necessary to provide suitable environmental conditions for plants to grow. In higher plants, the roots play an important role in their growth since the uptake of all the nutrition, including water, is done through the roots (Bibicova and Girlroy, 2003). This indicates that the control of the chemistry of the environment around the roots is the key factor for plants to grow strongly. Recently, electrochemical signal detection for the higher plants around the roots has been investigated. Iwabuchi *et al.* (1989) observed the electric patterns around growing cress roots. They reported that a change in the electric patterns was brought about by the growth of the root in a given environment.

Toko and his coworkers reported on the occurrence of the current flow around plant roots (Ezaki *et al.*, 1988), the current flow picture outside the root being shown to be related to growth. Miwa and Kushihashi (1922) reported on the stereoscopic electric current density picture around the root. Their current flow picture was constructed on the basis of the measured spatial assumption of the presence of an active ionic flow. They expected hormone accumulation in the region of growing portion of the root. However, neither the ionic concentration profile, nor the time dependence of the potential profile appears to have been studied in detail for the root surface during growth. It is interesting to measure directly the ionic concentration profile during the growth.

Electrochemistry can afford to control the electric field near the plant root. If the electrochemical potential is applied to the root, the signal is supposed to be sensed near the inferior of the root, which may possibly stimulate the growth of the plant.

It was found that plant growth was accelerated by the application of DC or square wave voltage to the root of a bean radicle planted in a culture bath (Mizuguchi *et al.,* 1994). It is worthwhile to study how the growing intact plant root is related to the presence of the substance which originates the potential distribution around the root surface. The ATP cycle is suggested to take part in the change of growth rate. This implies that the plant growth is influenced dynamically, not only by the concentration of the chemical entities, but also by the profile of electrochemical potential distribution in connection with ATP cycles.

In this context, the article of Tsutomu Takamura (2006) will introduce his work at first on the relation of potential distribution and ionic entities around the root, specially paying attention to the proton concentration followed by showing proton concentration image around the root during the growth and how the proton concentration affects the growth, leading to effective potential application to the root for accelerating the growth, and finally the investigation of the mechanism of the growth acceleration.

Environmental Stimulation and Adaptation of Plants

Plants have evolved sophisticated systems to sense environmental stimuli for adaptation and signals from other cells for coordinated action.

Consequently plant generates intracellular and intercellular electrical signals in response to these environmental changes (Bertholon, 1783; Bose, 1907; Bose, 1913; Bose, 1918; Bose, 1926; Ritter, 1811; Takamura, 2006). Neurotransmitter like compounds, such as acetylcholine, dopamine, histamine, noradrenaline and serotonin, participate in the information processes in plants (Roshchina, 2001).

The excitation reaction may travel between the top of the stem and the root in either direction. Plant electro-stimulation can have influence on the growth of plants (Takamura,2006; Biswas,2002). Mizuguchi etal(1994) applied one volt to the cultured solution and it accelerated by 30 per cent growth of bean sprouts.

Electrostimulation by a sinusoidal wave from a function generator induces electrical response in leaves with a phase shift. If voltage gated ion channels are closed and not involved in signal transduction along a plasma membrane, the propagation of passive electronic potentials can be described by a cable theory along a circuit consisting of plasma membrane capacitors C_1 and resistors R_1 and resistance along a plasma membrane R_2.

Electrical circuits in the roots and at the root/soil interface are very complicated and many authors proposed different active and passive equivalent electrical schemes. Due to additional Re-circuits in a root and soil, the duration of electronic potentials can increase.

Electrical Stimulations to Plants

The cells of many biological organs generate electric potential that results in the flow of electric current.

Electrostimulation of plants can induce activation of ion channels and ion transport, gene expression, activation of enzymatic systems electrical signaling, plants movements, enhanced wound healing, plant cell damage and influence plant growth. The electrostimulation by bipolar sinusoidal or triangular periodic waves induces electrical responses in plants, fruits, roots, and seeds with fingerprints of generic memristors.

Propagation of electrical signals inside plants can induce electrical signals between surface electrodes attached to a leaf. In many publications, the external refference Ag/AgCl electrode was inserted in the soil, compost, or aqueous buffering solution. Relationships between results, such as, amplitude, duration and speed of electrical signal propagation are obtained using inserted electrodes, attached to a leaf or external electrodes in a soil are unknown.

The work of Biswas (2002) on the effect of applied electric field on germination and growth of plants seedlings in Indian context has been found to be beneficial in respect of percentage of germination and early development.

A series of experiments have been conducted by him between 1.11.96 to 22.5.98 with seeds of gram (*Cicer arietinum*); jute (*Chorchorus olitorius*); paddy (*Oryza sativa*); raddish (*Raphanus sativus*); peas (*Pisum sativum*); and nate (*Amaranthus viridis*); where electric field of different types, strength and duration have been applied on the seeds of above mentioned species during their germination and early growth stages.

He used electrically insulated glass pots and rectangular plastic troughs, where finely powdered garden soil was used as germinating and growing media. Seeds have been embedded in rows near positive and negative electrodes and also in the mid-field between the electrodes. 18 gauge aluminium sheets were cut to shape and inserted into the soil at the opposite extremities of the glass pots and plastic troughs, touching the bottom of the containers, which acted as positive and negative electrodes carrying electric current and producing a homogeneous electric field through the moist soil column, when energized. Distance between the electrodes in case of glass pot and plastic troughs were kept at 12cm and 24cm respectively.

Terminals of 9 and 18 volts DC output was connected to the electrodes through an ammeter in the circuit, so as to know the amount of current drained during exposure. A voltmeter was also connected, parallely to measure the potential difference of the applied current. The field strength of current flow was measured by the probe of one square cm by inserting it into the soil, mid-field between anode and cathode. While using impulse current an electronic pulser producing one to four pulses per second was introduced into the circuit. The electrical resistance of the soil was measured by an ohm-meter.

A minimum of 30 and maximum of 100 seeds of each varieties were used for each set of experiment. A control group of equal number of seeds under identical conditions was maintained for comparison.

While using DC and impulsed DC of 9 and 18 volts for each set of experiment, intermittent current exposure, that is; current exposure of 1 to 8 hours per day for 3 to 8 days and their cumulative effect was observed in different sets of experiments. Continuous impulse current of 1/sec for 96 hours in 4 days were also applied as 9 and 18 volts in two sets of experiments along with control.

Responses to Applied Electric Current

Nine volt DC, when applied through moist garden soil having an electrical resistance of 4000-9000 ohm/cm^2, through vertically placed electrodes, an uniform electrical field, where the density of current flow to the tune of 34 to 200 µA/cm^2 and potential difference of 150 to 750 mV/cm was created surrounding the implanted seeds of *Cicer arietinum, Chorchorus olitorius, Oryza sativa, Raphanus sativus, Pisum sativum* and *Amaranthus viridis*, 83 per cent of *C. arietinum*, germinated after 15 hours of current exposure in 6 days as against 59 per cent sprouting in control group. Germination of *C. olitorius* and *O. sativa* took place after 158 hours of electric treatment in 7 days. Their sprouting percentages were 44 and 50 as against 52 and 20 percent of control groups respectively. After 24 hours exposure in two days, seeds of *R. sativus* and *P. sativum* germinated in 66 per cent and 80 per cent cases; whereas, 62 per cent and 40 per cent sprouting were noticed in their control group. Ninety percent germination was noticed after 36 hours of electric exposure in 3 days in respect of *A. viridis* as against 89 per cent germination in control group.

Table 1: Effect of DC Field on the Germination and Growth of different Variety of Seeds

| Type of Seeds | Type and Strength of Current - DC, 9 Volt | | | | Resistance of Soil in Ohms - 4000-9000 ohm/cm² | | | | | |
| | Current Intensity Surrounding Seeds | | Per cent of Germination at the Start | | Total Current Exposure | | Per cent of Survival and Growth at the End of Exposure | | | |
	mV/cm	µA/cm²	Control	Expt.	Hours	Day	Control	Near +ve	Near -ve	Mid - field
C. arietinum	150 - 750	50-200	59	83	15	6	9	30	10	20
C. olitorius	430 - 600	-	52	44	158	7	60	100	44	68
O. Sativa	430 - 530	-	20	50	158	7	73	90	90	90
R. sativus	-	34-120	62	66	24	5	62	66	-	-
P. sativum	-	34-120	40	80	24	5	40	80		
A. viridis	350 - 700	-	89	90	36	3	89	90		

Atmospheric temperature -28 - 30°C.

At the end of 15 hours exposure in 6 days, 60 per cent treated *C. arietinum* seedlings grew to 8 to 15 cm in length with 2 to 3 branches as against 9 per cent, 10 to 11 cm plants without any branches in control group. It was further observed that 30 per cent seedlings having 3 branches each and between 11-15 cm length grew near positive electrode. Only 10 per cent with 2 branches of 8-9 cm length were alive near negative electrode. In midfield, 20 per cent seedlings of 12-13 cm length with two branches survived at the end of 6 days.

After 158 hours current treatment for 7 days, the seedlings of *C. olitorius* survived in 63 per cent cases in treated ones, against 60 per cent in control group; whereas 90 per cent survival was noticed in seedlings of *O. sativa* in electrically treated ones as against 73 per cent in their control counterpart. Among growing seedlings of *C. olitorius*, 100 per cent of them near positive electrode were found to be growing to a length of 1.0 to 3.6 cm as against 44 per cent (1.1-2.5 cm length) near negative electrode and 68 per cent (2.4-4.7 cm length) in mid-field between two electrodes. In case of *O. sativa* however, the survival percentage was 90 per cent as against 73 per cent in control group. Equal numbers (90 per cent) of sprouting were noticed in all three positions, that is, near positive, near negative, and in between these two electrodes. Their growth were 2.7 to 5.5 cm near anode, 1.1 to 7.0 cm near cathode and 2.1 to 8.0 cm in the mid-field. Seedlings of *R. sativas* after 5 days with 24 hours current exposure grew to 1.1 to 3.4 cm while in control group, their growth were between 3.0 to 10.9 cm. Their percentage of survival being 66 per cent in treated cases and 62 per cent in control group. 80 per cent survival and growth of seedlings to 0.7 to 1.4cm was noticed in case of *P. sativum* as against 40 per cent survival and 1.0 to 1.1 cm length in control group in identical condition of treatment as that of *R. sativus*. Seeds of *A. viridis* when treated with electric current for 190 hours in 8 days, the germination was 90 per cent as against 89 per cent in untreated group. Their growth, however, were equal in the range of 0.2 to 3.3 cm in both experimental and control group.

Seeds of *C. arietinum* when exposed to 9 volt DC field for 6 hours in 3 days, germinated in 83 per cent cases as against 59 per cent in control group. After 6 days with 15 hours of treatment, the percentage of germination and growth were were 60 and 9 percent in experimental and control group respectively. Seedlings in experimental group exhibited 2 to 3 branching of 8 to 15 cm length of which 30 per cent grew near anode, 10 per cent near cathode and 20 per cent were in mid-field. In another set, 100 per cent germination of *C. arietinum* was noticed after electric treatment of 22 hours in 7 days as against 17 per cent in control group. After 24 days with 44 hours of electric exposure, 40 per cent seedlings with two branches of 28-32 cm length survived as against 7 per cent in control group, which attained a length of 25-39 cm without any branching. Out of these 40 per cent in experimental group, 30 per cent were near the anode and 10 per cent in the mid-field.

When treated with impulse DC of 9 volts (1/sec), equal number of seeds (50 per cent) sprouted both in control and experimental group after 96 hours of electric exposure in 4 days. At the end of 190 hours treatment in 8 days, the survival

Table 2: Effect of different Type of Seed – *C. arietinum* and Strength of Electric Field on Seed Germination and Growth

Input Voltage and Nature of Current	Current Intensity of the Field		Percentage of Germination at the Start		Total Current Exposure		Percentage of Survival and Growth at the End of Electric Exposure			
	mV/cm	$\mu A/cm^2$	Control	Expt.	Hours	Days	Control	Near +ve	Near -ve	Mid-field
9 volt DC	100 - 750	50 - 200	59	83	15	6	9	30	10	20
9 volt DC	100 - 900	40 - 400	17	100	44	24	7	30	-	10
9 volt pulsed Dc1/sec	350 - 700	-	50	50	190	8	85	30	20	-
18 volt pulsed DC1/sec	220 - 540	-	50	30	190	8	85	20	-	-

Atmospheric temperature -28 - 30ºC.

Resistance of soil in ohms – 4000 – 8000/cm².

percentage was 85 per cent in control group and 50 per cent in experimental group of which 30 per cent grew to 45 – 70 cm in length near the positive electrode and 20 per cent attained a length of 17 – 52 cm near the negative electrode.

At 18 volts impulse DC (1/sec) after 4 days with 96 hours treatment only 30 per cent germinated, while in control group germination was 50 per cent. At the end of 190 hours electric exposure in 8 days, 20 per cent seedlings of 22-59 cm length survived near the anode, while in control group 85 per cent seedlings survived and grew to 34-72 cm length.

Plants reported to have responded to most kinds of music or sound, to magnetic and electric field or currents and all of which favored their growth under certain conditions (Van Mater, 1987).

Exposure to low voltage DC field (9 volts) on germinating seeds of various species in the work of Biswas (2002), were marked with varied effects on the percentage of germination and growth of seedlings in the beginning and after prolonged treatment. Though the electric treatment could not shorten the incubation period, but a marked difference in the percentage of germination at the beginning and subsequent development was noticed among different species. Except *C. olitorius*, the percentage of germination in *C. arietinum, O. sativa, R. sativus, P. sativum and A. viridis* were more in electrically treated groups. At the end of prolonged treatment of 5 to 8 days, with 15 to 190hours of electric current exposure, the percentage of survival was more in experimental groups and particularly, near the positive electrode, indicating a beneficial effect of electric field on plant growth near the anode. Next to the anode, the percentage of survival was more in mid-field than near the negative electrode which indicated that the condition near the cathode was not congenial for the growth of seedlings.

Branching of *C. arietinum* seedlings (2-3 branches) during current exposure was another interesting phenomenon, which never occurred any of the untreated group during that period, may be due to the effect of electrical stimulation on growing seedlings in early stages.

De Meo (1985) found six fold increase in the length of mung bean sprouts inside a strong accumulator, as compared to a control group of sprouts. He claimed that the germination rates, budding, flowering and fruiting can be increased by charging seeds or growing plants directly inside the accumulator.

Treating seeds of *C. arietinum* with pulsed DC (1/sec) both in low (9 volts) and high voltage (18 volts), the control group germinated similarly or even better than electrically treated group, both during initial and prolonged treatment. At higher voltage, the survival of experimental group decreased further, showing that impulse current was not favourable.

It also failed to form branching of seedlings, though in continuous DC of similar current strength, branch formation was noticed within 6 days at 15 hours current exposure, in every cases.

Sir J. C. Bose, cited by Scheminzky *et al.* (1938) devised a way for plant to know its reaction to electrical stimuli. He also found that in the cortex tissue of the growing layer, plants have pulsating heart cells that help pump the sap up through the stem.

All these earlier findings and the results obtained from the work of Biswas (2002) leads to show that plant, like many other stimuli, do respond to the electrical stimuli and their early development process is influenced by such stimulation can become favourable under certain conditions.

Self Induced Galvanic Current

Galvanic current is produced in the presence of two or more similar or dissimilar metals in an electrolyte (including freshwater) and salt water environment. Salt water and acidic solutions are good electrolyte. Freshwater is a poor electrolyte. It is a type of electrical energy that can be measured. Galvanic current is also defined as the unidirectional current of an electric charge. It may flow through conductor materials like, wire but also capable of travelling in insulators and semi-conductors or even via a vacuum like ion beams. Galvanic corrosion is caused by self induced current.

Experiment on induced galvanic current started from June, 2016 in the electrophysiological laboratory, under the leadership of Dr. K. P. Biswas to observe the effect on blood worm (*Tubifex* sp.).

Preliminary experiments have been conducted in order to ascertain the nature of galvanic current and the electric field that was created when two similar and dissimilar metal plates (electrodes) were immersed in fresh water (weak electrolyte), and if the induced current thus produced is measurable in terms of current intensity (μA) and potential difference (mV) at different period of immersion of electrodes (metal plates).

It has been observed that induced galvanic current is produced between two similar and dissimilar electrodes in freshwater electrolyte and can be measured with mili-ammeter (for current intensity) and mili-voltmeter (for potential difference), though it is very small in quantity. Values of current intensity and potential difference were at the peak immediately after immersion of the electrodes into the water (electrolyte) and connecting the electrodes either to a mili-ammeter or a mili-voltmeter.

Nature of Current Flow		Pattern of Current Flow	Polarity	Duration (Hours)	Dates	Electrodes/ Distance between Electrodes	Electrolyte	Resistance/ Conductivity of Electrolyte	Temperature of Electrolyte (^{o}C)
Start	End								
400mV	20 mV	400 → 0	Normal	34	13/07/18 to 14/07/18	Aluminium (28×26cm)	Fresh water	2800 Ω/cm^2	30.5
		0-10	Reverse			28 cm		200 µS/cm	
		10-50	,						
		50-28	,						
		28-40	,						
		40-20	,						
		20-35	,						
		35-20	,						

| Nature of Current Flow | | Pattern of Current Flow | Polarity | Duration (Hours) | Dates | Electrodes/ Distance between Electrodes | Electrolyte | Resistance/ Conductivity of Electrolyte | Temperature of Electrolyte (°C) |
Start	End								
230mV	0 mV	230-20	Reverse	76	29/07/18 to 1/08/18	Aluminium (28×26cm)	Fresh water	5600 Ω/cm²	28
		20-0	,						
		0-40	Normal			28 cm			
		40-100	,						
		100-50	,						
		50-0	,						
		0-20	Reverse						
		20-100	,						
		100-70	,						
		70-200	,						
		200-120	,						
		120-22	,						
		22-0	,						
		0-400	,						
		400-115	,						
		115-70	,						
		70-3	,						
200 mV	20mV	200-15	Normal	24	04/08/18 to 05/08/18	Aluminium (28×26cm)	,	3000 Ω/cm²	30
		15-35	,			119cm		30 µS/cm	
		35-20	,						

The observations were recorded at intervals, continuously for 72 hours. The insertion of mili-ammeter or mili-voltmeter in the circuit resulted connecting a load (depending on the internal resistance of the meter) on this low amount of induced galvanic current produced in poor electrolyte (fresh water). As a result the current strength and potential difference from a peak value rapidly decline witin a minute and thereafter either stabilize or very slowly decline for hours together.

With two aluminium plates (electrodes) 0f 28×26cm, when separated at a distance of 28cm in freshwater with an electrical resistivity of 2800 Ω/cm^2 (conductivity - 200 µS/cm) and water temperature of 30.5°C, the potential difference between the electrodes registered 400 mV (normal polarity) in one case, which dropped down to 0 mV within 13 minutes (780 seconds). The observation was continued for 34 hours. The potential difference from 0 mV increased to 10 mV and thereafter to 50 mV and began to oscillate from 50 mV to 28 mV; 28 mV to 40 mV; 40 mV to 20 mV; 20 mV to 35 mV; and 35 mV to 20 mV, till the end of observations after 34 hours, all in reverse polarity.

In the second test, under the same conditions, the initial potential differences from 230 mV between the electrodes dropped down to 20 mV within 5 hours (Reverse polarity). The galvanic current resumed the normal polarity from reverse polarity (20 mV) of 40 mV to 100 mV within 4 hours and from 100 mV to 50 mV within 2 hours. Again the polarity of current changed to reverse direction and registered 20 mV to 100 mV in 2 hours; 70 mV to 200 mV in 10 minutes, 120 mV to 22 mV in 12 hours, 22 mV to 0 mV in 5 hours, and finally 70 mV to 3 mV in 19 hours, all in reverse polarity.

Using the same size of aluminium electrode (28×26cm) but separated at a larger distance of 119 cm in freshwater (24 cm depth) having an electrical resistance of 3000 Ω/cm^2 (water conductivity - 30 µS/cm), the potential difference was 200 mV at the start, which gradually declined to 15mV within 11 hours. From 15 mV, the potential difference again increased to 35 mV in 45 minutes and from 35 mV again declined to 20 mV in 6 hours, all in reverse polarity.

In the above experiments, the oscillation of galvanic current (mV) from higher value to lower and again from lower to higher value and finally from higher to lower value during the period of observation may be due to continuous load (internal resistance of millivoltmeter) without any break during the period of observations. It is not known if the constant load application without break can also cause a change in polarity.

In view of this supposition, next set of experiments were conducted with an interruption of applying load (*i.e.* disconnecting the mV meter and again connecting the meter after some time).

The same electrode (aluminium plate of 28×26cm),separated at a distance of 28cm was used in freshwater at 30°C, the electrical resistance being 2800 Ω/cm^2. The disconnection of the load (mV meter) was 5 to 8 minutes in the beginning, thereafter a long break of 150 minutes followed by a break of 5 minutes to one minute. The peak and lowest potential difference was measured at the start and after one minute of continuous current flow.

Highest and lowest potential difference of 230 mV and 15 mV was observed in the beginning. The average of potential difference (mV) after each break and after 1 minute of current flow at different period of interruptions were;

- ☆ 150 mV to 15 mV in 8 minutes break
- ☆ 100 mV to 11 mV in 5 minutes break
- ☆ 132 mV to 12 mV in 5 minutes break
- ☆ 110 mV to 10 mV in 4 minutes break
- ☆ 90 mV to 10mV in 3 minutes break
- ☆ 60 mV to 9 mV in 2 minutes break
- ☆ 30 mV to 6 mV in 1 minutes break
- ☆ 165 mV to 10 mV in 15 minutes break (Reverse polarity)
- ☆ 60 mV to 5 mV in 15 minutes break (Normal polarity)
- ☆ 95 mV to 9 mV in 10 minutes break (Normal polarity)
- ☆ 68 mV to 6 mV in 5 minutes break

Earlier all the observations on potential difference (mV) were measured by connecting the electrodes (Aluminium plates) to milli-voltmeter which gave rise several fluctuations in the mV. To find out a steady exact quantity of galvanic current, observations were made on the current intensity (µA) and potential differences (mV) independently in the circuit and from the field. As such, three sets of measurments were taken; (a) between electrode, (b) between positive and a probe electrode of 1 cm^2, and (c) independently from field with the help of probe electrodes of one square cm, separated at a distance of 1 cm.

28×26cm aluminium plates were used as electrodes, which were immersed in fresh water, the electrical resistance of which 3000 Ω/cm^2 at 29^{0}C. The electrodes were separated at a distance of 28cm. Three sets of breaks of 30 minutes, 60 minutes and 120 minutes were taken, the average of which were;

At the start with 30 mins break, there was a progressive decrease from 120 mV to 100 mV to 16 mV, which after stoppage of 1 minute of current flow again decreased from 80 mV to 70 mV to 40 mV.

The current intensity after 30 minutes break registered 110 µA which decreased 105 µA to 60 µA to 25 µA and after stoppage of 1 minute of current flow from 6 µA to 5 µA to 1 µA.

With 60 minutes break, the average was 70 mV to 60 mV and thereafter from 48 mV to 46 mV after 1 minute stoppage of current flow. The current intensity after 60 minutes interruption was 70 µA to 65 µA and remained constant to 7 µA after 1 minute.

After 120 minutes interval the potential difference was 70 mV in the beginning and in all cases were 46 mV after 1 minute. The current intensity after 120 minutes break was 74 µA to 73 µA to 70 µA in the beginning and stabilized to 8 µA to 7 µA after 1 minute.

It has been found that 120 mins interval could stabilize the generation of galvanic current without much fluctuations (70 mV to 46 mV). A moderate fluctuations were noticed in 60 mins interruption (70 mV to 60 mV and 48 mV to 46 mV). Highest fluctuations were noticed at 30 mins interval (120 mV to 16 mV and 80 mV to 10 mV).

The curent intensity (µA) followed the same pattern to that of potential difference (mV). Therefore 120 interval (break) was found to be the best followed by 60 mins breaks with regards to production of self induced galvanic current.

With regards to the procedure to be adopted for taking measurements of (120 – 16 mV and 10 µA), the probe of 1 cm^2 and positive electrode gave the magnified value of 820 to 700 mV (540-720 mV) and 550 to 850 µA (80-130µA) followed by between electrodes (12-16 mV and 80-10 mV) and (110-25µA and 8-5 µA) and the lowest value of 1 cm^2 probes (102-48mV and 20-52 mV) and (7-10 µA and 3-6 µA). Therefore, values between the electrodes were considered as values nearest to the actuals.

Electrodes of similar and dissimilar metals were tested with aluminium and zinc plates. Electrodes of 27.8×6.8 cm in both the cases separated at a distance of 28 cm in freshwater were tried for the generation of induced galvanic current.

Aluminium + Aluminium produced 350 mV reduced to 20 mV in 30 minutes (RP)

Aluminium + Zinc produced 350 mV to 60 mV for 21 minutes (RP)

         600 mV to 90 mV for 40 minutes (RP)

         65 mV to 120 mV for 500 minutes (RP)

         120 mV to 125 mV for 735 minutes (RP)

Zinc + Zinc produced 700-145 mV to 30-90 mV for 80 minutes (RP)

         50 mV to 20 mV for 100 minutes (NP)

         0 mV to 220 mV for 660 minutes (RP)

         60 mV to 0 mV for 960 minutes (NP)

Dissimilar electrodes like aluminium and zinc maintain a steady flow of induced galvanic current than two similar electrodes of Zn and Al.

Field Application of Self Induced Galvanic Current: Seeds and Seedlings

The first experiment of application on self induced galvanic current on the germination of Bengal gram seed was done in a petridish with silver electrodes, through a water media, having a temperature of 30°C. Ten gram seeds, in each experimental and control petridish were used for the test. The voltage and current intensity, thus produced by silver electrodes fluctuated from 160 mV to 260 mV and 21 μA to 32 μA during 15 hours, when 100 per cent germination with 1cm root formation was noticed in experimental petridish. During the period 60 per cent germination with more than 1cm root formation was effected in control petridish.

Changing over to glass trough with aluminium electrodes on the third day in the same watery media, the potential difference and current intensity registered to 470 – 500 mV and 208 – 300 μA respectively. At that time in experimental trough 4 shoots of 1 – 1.2 cm, 3 shoots of 2 – 3 mm were visible. Roots (2 – 3 cm) of all the seedlings went down. At the same time in control, 3 shoots (0.5 – 0.8 cm) became upright and roots (1 – 2 cm) of 6 seedlings moving down.

On the fifth day under the influence of 500 – 520 mV and 400 – 800 μA and in a changed culture media, (replacing water with 2:1, soil extract: tea leaf extract) 9 seedlings in experimental trough grew to 1.2 – 6.7 cm (shoot of the remaining seedling coming up). Main roots of all the 10 seedlings attained a length of 1 – 8 cm, showing branching of roots (2 – 4 mm length) in 6 seedlings. In the control, at that time 5 seedlings grew to 1.5 - 6.2 cm, two germinating, with 1.2 – 5.8 cm roots in 6 seedlings. Branching of root was not visible then.

On the eighth day under the influence of 530 – 540 mV and 100 – 1000 µA, all the ten seedlings grew in three range; 6.1 – 6.5 cm (2 seedlings); 11.2 -11.5 cm (2 seedlings); 16.8 – 17.8 cm (6 seedlings) in experimental conditions. While in control trough 7 seedlings grew in three range; 7 – 10.2 cm (2 seedlings); 14.8 – 15.6 cm (2 seedlings); 17.4 to 18.8 cm (3 seedlings). They have been provided with 170 ml of soil extract (130ml) and tea leaf extract (40 ml) every alternative day.

On 12th day under the influence of 550 – 560 mV and 100 – 1000 µA of induced galvanic current generated by the immersed electrodes, all the 10 seedlings of the experimental trough attained a length of 20.5 – 32.5 cm; while in control trough 7 seedlings grew to 27.2 – 28.4 cm during that period.

On the 21st day under the influence of same induced galvanic current the gram seedlings exhibited the following growth of shoots and roots along with branching of roots.

Experiment

Length of shoot (cm)	38.2, 37.5, 34.2, 38.5, 38.5, 37.5, 27, 20, 34.5, 30
Length of root (cm)	8, 7.5, 9.4, 8.2, 9, 3.5, 7.5, 8, 7.5, 8
No. of brands in the roots	13 (2 – 3 cm), 8 (2.5 – 3 cm), 7 (1 – 2.5 cm), 13 (1 – 3.5 cm), 15 (1 – 3.5 cm), 7 (1 – 2.6 cm), 10 (1 – 2.5 cm), 18 (3.5 – 4.5 cm), 19 (1.5 – 3 cm), 35 (1.5 – 4 cm)

Control

Length of shoot (cm)	38.5, 41.8, 37.0, 38.4, 33.5, 38.5, 37.5
Length of root (cm)	11.2, 9.0, 8.7, 9.5, 9.0, 8.5, 6.5
No. of brands in the roots	8 (3 – 4.5 cm), 26 (3.5 – 6 cm), 16 (3.5 – 5.8 cm), 18 (3.5 – 5.8 cm), 18 (1 – 6.5 cm), 15 (1 – 5 cm), 19 (1 – 4.5 cm)

The profuse branching of roots in all the seedlings (both experimental and control troughs) indicated the scarcity of nutrients in the liquid media, for which the seedlings without branching of shoots (no branching of shoots observed) rapidly radiated by branching of roots to fight the adequacy of nutrient in culture media.

It was therefore, thought to transfer the seedlings either to the soil media with or without galvanic current coverage or to sow the seeds directly in the soil.

The beneficial effect of induced galvanic current was 100 per cent germination and more branching (7 - 35) as against 70 per cent germination and 8 – 26 branches in control.

Soil as Growing Media

Observing the stunted growth in nutrient deficient watery media, steps were taken to conduct experiments in soil media in electrically insulated plastic trays. In this series the depth of soil was maintained at 2 cm initially in 28 X 23 cm tray, which subsequently increased to 4 cm and 4.5 cm in the following series. Two experiments (Exp – I and Exp - 2) with galvanic current stimulation through Aluminium and Zinc electrodes and one control without electrical stimulation were first taken up. In each of two experimental trays and in control, 10 sunflower (*Helianthus annuus*); 10 chickpea (*Cicer arietinum*); 10 cucumber (*Cucumis sativus*); 10 snake bean (*Vigna unguiculata*) and 2 snake gourd (*Trichosanthes cucumerina*) seeds were embedded into soil with adequate watering of the soil.

Galvanic current (generated by the electrodes), characteristics of two experimental trays were measured and recorded twice a day and after every 12 hours with respect to potential difference (mV) and current intensity (µA) is given in a graphical form. In the experiment I, the potential difference fluctuated from 260 – 355 mV and the current intensity from 55 to 125 µA during these 10 days; while in experiment II the potential difference varied from 190 to 320 mV and the current intensity from 40 to 98 µA (Table 1, Figure 1).

Table 1: Measurements of Voltage and Current Intensity (At 24 hrs interval)

Date	Exp - I		Exp - II		Date	Exp - I		Exp - II	
	mV	*µA*	*mV*	*µA*		*mV*	*µA*	*mV*	*µA*
31.03.19	310	100	260	70	06.04.19	320	93	220	60
01.04.19	343	102	295	98	07.04.19	305	75	255	69
02.04.19	352	125	280	84	08.04.19	347	95	320	73
03.04.19	351	122	225	70	09.04.19	260	60	190	46
04.04.19	355	115	255	59	10.04.19	270	55	190	40
05.04.19	347	111	252	63					

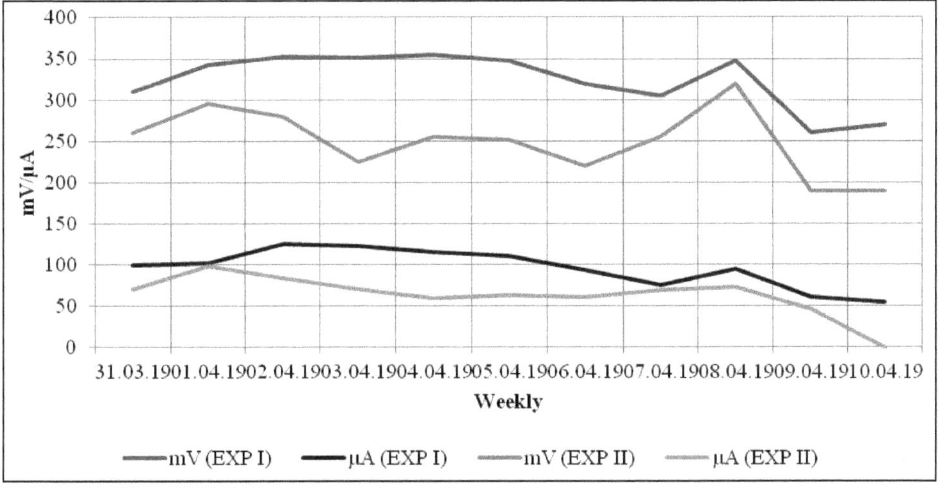

Figure 1: Pattern of Galvanic Current Intensity around Plant Seed and Seedlings.

The percentage of germination after 120 hours of putting seeds into the soil were;

Sunflower – Control – 30 per cent

Exp – I – 40 per cent

Cucumber – Control – 50 per cent

Exp – I – 70 per cent

Snake bean – Control – 30 per cent

Exp – I – 30 per cent

Exp – II – 20 per cent

Chickpea and snake gourd did not germinate at all both in experimental and control trays.

Three days after germination, the growth of seedlings were measured for length and the increment in growth calculated.

In control 30 per cent of germinated sunflower grew from 1.5 cm to 5 cm, an increment of 3.5 cm in 3 days (1.167 cm/day).

In Exp I 40 per cent of germinated sunflower had grown from 1 cm to 5.7 cm, an increment of 4.7 cm in 3 days (1.57 cm/day).

50 per cent cucumber, in control grew from 2.6 – 4.5 cm to 6.7 – 8.2 cm in 3 days, an increment of 4.1 – 3.7 cm (1.3 cm/day).

In Exp – I; 70 per cent cucumber grew from 5.2 cm to 6.8 cm, an increment of 1.6 cm within 3 days (0.53 cm/day).

In control, snake bean grew from 4.6 – 5.0 cm to 9 – 9.5 cm in length, an increment of 4.4 to 4.5 cm within 3 days (1.48/day).

In Exp – I 30 per cent snake bean grew from 4.2 cm – 8.5 cm to 9.7cm and 10.2 cm, an increment of 5.5 – 1.7 cm within 3 days (1.2 cm/day); while in Exp – II 20 per cent seedlings (snake bean) have grown from 1.5 cm to 2.7 cm, an increment of 1.2 cm in 3 days (0.4 cm/day).

But if overall growth (length) of the seedlings is considered, then, it has been observed, that sunflower seedlings in experimental trough attained a length of 2.5 – 5.7 cm as against 3.7 – 5.0 cm in control on 8th day after putting seeds into the soil. Snake bean registered a length of 9.7 to 10.2 cm in experimental conditions, as against 3.3 – 9.5 cm in control conditions. Cucumber, however, grew well without galvanic current influence (in control 6.7 – 8.2 cm as against 5.2 – 6.8 cm in experimental conditions.)

Due to storm and heavy rains many of the seedlings were uprooted and damaged, resulting to stoppage of the experiments.

However marginal effectiveness of galvanic stimulation was observed in percentage of germination in sunflower and cucumber and overall growth in sunflower and snake bean.

Experiments were repeated in the same plastic trays with 4 cm soil and zinc and aluminium electrodes. The electrical resistance of the soil was 1150 ohms/cm^2 in ExptL – I tray and 2000 ohms/cm^2 in ExptL – II tray. The soil of control tray was having an electrical resistance of 1500 ohms/cm^2.

Following seeds have been put in each of the three trays:

- ☆ Sunflower (*Helianthus annuus*) – 10
- ☆ Bengal gram (*Cicer arietinum*) – 10
- ☆ Cucumber (*Cucumis sativus*) – 3
- ☆ Snake bean (*Vigna unguiculata*) – 5
- ☆ Ladies finger (*Abelmoschus esculentus*) – 5
- ☆ Luffa (*Luffa acutangula*) – 5

A potential difference of 4000 mV in ExptL – I and 3600 mV in ExptL – II and a current intensity of 250 µA in ExptL – I and 150 µA in ExptL – II were recorded at the time of putting seeds in experimental trays. Experiments in the series continued for a month from 10.4.19 to 9.5.19. During this period, the potential difference in ExptL – I varied from 206 to 533 mV; while in ExptL – II, the values ranged from 22 mV to 310 mV. The current intensity of ExptL – I, during this period fluctuated between 38µA to 200 µA and in ExptL –II between 7 µA to 118 µA (Table 2, Figure 2).

Table 2: Measurements of Voltage and Current Intensity (At 24 hrs interval)

Date	Exp - I		Exp - II		Date	Exp - I		Exp - II	
	mV	µA	mV	µA		mV	µA	mV	µA
10.04.19	314	84	222	60	25.04.19	435	130	215	93
11.04.19	378	134	274	102	26.04.19	450	38	168	70
12.04.19	432	149	310	110	27.04.19	370	95	190	87
13.04.19	454	190	252	110	28.04.19	390	101	130	95
14.04.19	405	123	274	103	29.04.19	310	58	161	66
15.04.19	370	105	190	70	30.04.19	290	66	252	70
16.04.19	407	149	190	77	01.05.19	320	81	176	46
17.04.19	446	156	237	89	02.05.19	271	79	88	25
18.04.19	465	133	215	77	03.05.19	276	111	38	11
19.04.19	460	118	231	80	04.05.19	430	173	84	31
20.04.19	465	200	310	118	05.05.19	294	81	238	56
21.04.19	464	125	185	67	06.05.19	270	95	180	45
22.04.19	467	124	160	70	07.05.19	234	82	160	45
23.04.19	533	134	185	73	08.05.19	206	80	40	10
24.04.19	525	164	200	77	09.05.19	216	85	22	7

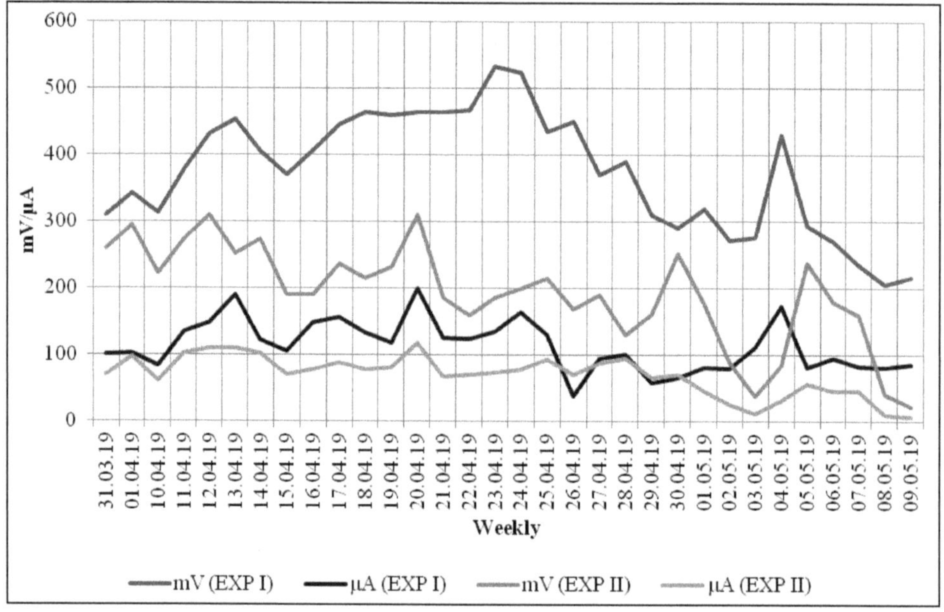

Figure 2: Pattern of Galvanic Current Intensity around Plant Seed and Seedlings.

After two days of putting seeds in experimental trays, the following seeds have been found to germinate in the experimental and control trays.

In ExptL – I – Sunflower 70 per cent; Cucumber – 66 per cent; Snake bean – 40 per cent and Luffa – 20 per cent;

ExptL –II – Sunflower – 10 per cent; Cucumber – 33 per cent;

Control – Sunflower – 30 per cent; Cucumber – 10 per cent; Snake bean – 10 per cent and Bengal gram – 10 per cent.

In five days and 7 days the seedlings registered a growth as under;

Seeds	*In 5 Days (in cm)*			*In 7 Days (in cm)*		
	Exp - I	*Exp - II*	*Control*	*Exp - I*	*Exp - II*	*Control*
Sunflower	1.3 - 6	5.6	2.1 -6.3	1.6 – 12.4	10.8	1 – 12
Cucumber	5.3 – 8.3	6.5	4.8 – 5.1	9.7 – 11.8	9.8	5.8 – 6.8
Snake bean	6.1 – 9.1	-	-	9.5 – 15.2	-	-
Luffa	6.9	-	-	9.2	-	-
Bengal gram	-	-	1.4	-	-	5.9

Except Cucumber, where growth in experimental trays was more than control, the seedlings of sunflower showed marginal differences between control and experimental troughs. Comparison could not be made with respect to snake bean and luffa as there was no germination of them in control. Only one Bengal gram seed germinated in control which grew to 1.4 cm in 5 days and 5.9 cm in 7 days. No Bengal gram seed germinated in experimental troughs and the seeds, after swollen with water have been found to be infested with ants and eaten by them.

Self induced galvanic current have been found to have beneficial effect on germination of sunflower, cucumber, snake bean and luffa by the way of percentage of increased germination. The comparative growth of cucumber seedlings were found to be more in galvanically influenced seedlings. In case of sunflower, however, the growth of seedlings were more in control than the galvanically influenced ones, which may possibly due to lower rate of germination in control by less than half of that of experimental I, having less number of seedlings in control facilitating increased growth at thinner population density.

Experiments (two at a time) have been conducted in the same troughs (28 X 23 cm) with 4 cm of soil in each including control. The electrical resistance of the soil (moist) of the germinating trays were measured at the beginning of onset of self induced galvanic current and was found to be 4000 Ω/cm^2 in ExptL – I, 2600 Ω/cm^2 in ExptL –II and 3500 Ω/cm^2 in control. Experimental troughs were provided with zinc and aluminium plate covering the entire cross – section of soil, breadth wise in the germinating tray for the coverage of all the seeds planted in the experimental troughs under the influence of galvanic current.

Seeds of the following types and numbers were planted in each of the three trays (two experimental and one control)

- ☆ Snake bean (*Vigna unguiculata*) – 10
- ☆ Sunflower (*Helianthus annuus*) – 5
- ☆ Luffa (*Luffa acutangula*) – 6
- ☆ Pumpkin (*Cucurbita pepo*) – 5
- ☆ Bottle gourd (*Lagenaria siceraria*) – 2

Experiments continued for 20 days from 13.6.19 to 3.7.19 when the highest peak of 4310 mV in ExptL – I was recorded on 27.6.19, the lowest peak was however 800 mV on 3.7.19. Peak current intensites of 175 µA, 183 µA were recorded in the same experimental trough (ExptL - I) on 18.6.19, 27.6.19 and 24.6.19 respectively. In the experimental trough no. II, highest potential difference and current intensites were observed on 24.6.19 as 4600 mV and 229 µA as current intensity. The lowest value of 1200 mV and 36 µA in ExptL – II was recorded on 3.7.19 (Table 3, Figure 3).

After two days of putting seeds in the soil media, germination of sunflower was noticed in 60 per cent in ExptL – I, 80 per cent in ExptL – II and 80 per cent in control. No other seeds germinated in any of the experimental and control trays.

Table 3: Measurements of Voltage and Current Dntensity (At 24 hrs interval)

Date	Exp - I		Exp - II		Date	Exp - I		Exp - II	
	mV	*µA*	*mV*	*µA*		*mV*	*µA*	*mV*	*µA*
13.06.19	2700	125	3300	110	24.06.19	3700	183	4600	229
14.06.19	1400	74	1470	64	25.06.19	4250	170	4500	188
15.06.19	2520	115	1780	65	26.06.19	4120	163	4050	160
16.06.19	2900	118	2520	80	27.06.19	4310	175	4150	178
17.06.19	3760	170	3170	124	28.06.19	2350	76	3450	141
18.06.19	3810	175	3200	133	29.06.19	2370	95	4080	183
19.06.19	2650	109	3550	145	30.06.19	2020	75	3170	125
20.06.19	1775	90	3500	125	01.07.19	1400	58	2250	85
21.06.19	2600	93	2910	120	02.07.19	2140	100	2300	95
22.06.19	1550	65	1900	77	03.07.19	800	24	1200	36
23.06.19	2700	123	3200	123					

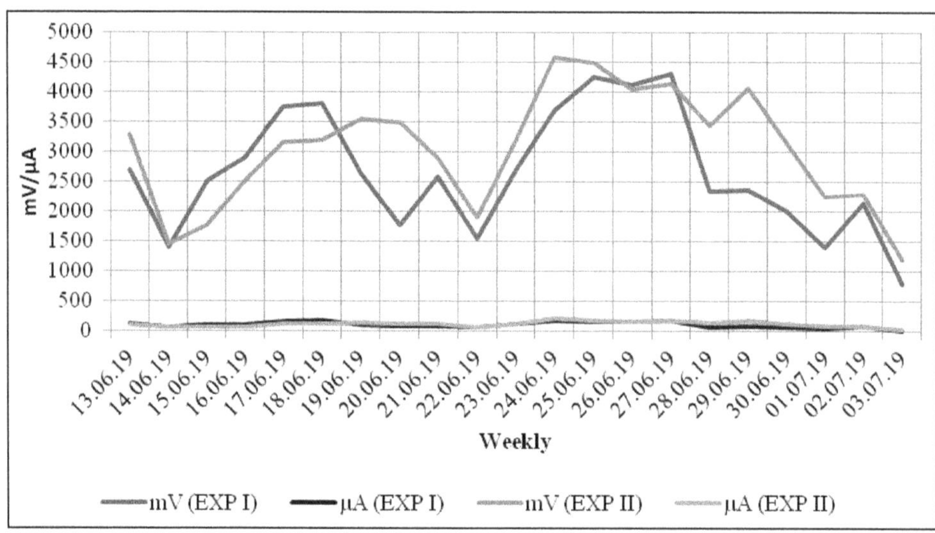

Figure 3: Pattern of Galvanic Current Intensity around Plant Seed and Seedlings.

After 12 days, sunflowers in ExptL – I grew, 13.4 – 20.6 cm, in ExptL – II from 11.4 to 18.9 cm and in control from 11.0 to 16.5 cm.

The increment in growth of sunflower in ExptL – I was calculated to be 1.44 cm/day; while that of ExptL – II and control were 1.33 cm/day ad 1.30 cm/day respectively.

The results of germination and incremental growth indicate that there is a positive influence of induced galvanic current on the percentage of germination and growth of seedlings under the coverage of galvanic current.

Experiments with two similar sets of electrodes, *e.g.* aluminium and aluminium (ExptL - I) and zinc and zinc (ExptL - II) were taken up in 28 cm X 23 cm plastic troughs with 4.5 cm deep soil along with control. Electrical resistance of soil being 4000 Ω/cm^2 in ExptL – I; 3600 Ω/cm^2 in ExptL – II and 4000 Ω/cm^2 in control.

Seeds of following varieties have been put in each of the three trays on 21.7.19,

★ Paddy (*Oryza sativa*) – 10

★ Cucumber (*Cucumis sativus*) – 10

★ Malabar spinach (*Basella alba*) – 10

★ Bitter gourd (*Monordica charantia*) – 2

Experiments continued for 28 days from 21.7.19 to 15.8.19, when the following variations of galvanic current coverage to the germinating seeds and its growth were observed in two experimental trays.

ExptL. tray – I – A maximum of 657 mV and 300 µA were recorded, while during the same period of minimum potential difference and current intensity were 71 mV and 0 µA respectively.

ExptL. tray – II – The highest potential difference of 662 mV and current intensity of 310 µA was observed as against the lowest of 154 mV and 40 µA.

The fluctuation of mV and µA recorded daily is indicated in Table 4 and Figure 4.

Table 4: Measurements of Voltage and Current Intensity (At 24 hrs interval)

Date	Exp - I		Exp - II		Date	Exp - I		Exp - II	
	mV	µA	mV	µA		mV	µA	mV	µA
21.07.19	320	300	399	190	03.08.19	256	60	182	45
22.07.19	313	140	385	140	04.08.19	320	75	186	40
23.07.19	504	270	525	230	05.08.19	500	175	320	70
24.07.19	479	205	554	235	06.08.19	188	50	212	40
25.07.19	488	145	613	190	07.08.19	280	20	169	45
26.07.19	470	145	553	190	08.08.19	221	5	245	55
27.07.19	330	110	297	84	09.08.19	242	10	330	110
28.07.19	315	100	250	75	10.08.19	181	15	494	135
29.07.19	200	75	278	70	11.08.19	230	5	283	80
30.07.19	283	80	276	70	12.08.19	188	5	178	40
31.07.19	393	120	457	125	13.08.19	71	0	188	40
01.08.19	657	215	662	190	14.08.19	123	5	278	310
02.08.19	379	110	459	125	15.08.19	214	43	154	125

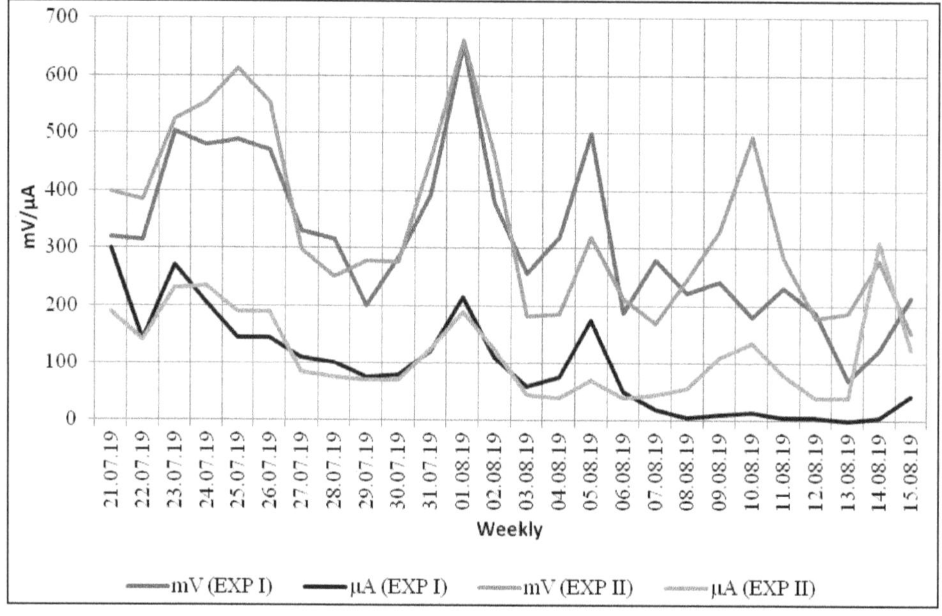

Figure 4: Pattern of Galvanic Current Intensity around Plant Seed and Seedlings.

As regards the germination of sunflower seeds, no germination of sunflower took place in ExptL – I and control. A total of 80 per cent sunflower seed germinated in ExptL – II in the following time schedule;

40 per cent germination after 48 hours of sowing

70 per cent germination after 96 hours of sowing

80 per cent germination after 168 hours of sowing

Paddy – 20 per cent germination in ExptL – II and control after 96 hours of sowing

Paddy – 50 per cent germination in ExptL – II and control after 168 hours of sowing

Paddy – 60 per cent germination in Control and control after 168 hours of sowing

Paddy – 40 per cent germination in ExptL – I and control after 168 hours of sowing

Malabar spinach – 10 per cent germination in ExptL – II after 96 hours of sowing

Malabar spinach – 10 per cent germination in ExptL – I after 168 hours of sowing

Cucumber – 10 per cent germination in ExptL – II after 168 hours of sowing

Growth Under the Influence of Galvanic Current

80 per cent sunflower germinated in ExptL – II and grew from 1.5 – 3.5 cm to 3.2 – 21.4 cm, an increment of 1.7 cm – 17.9 cm in 11 days.

50 per cent paddy germinated in ExptL – II and grew from 0.8 – 3.5 cm to 18.2 – 31 cm an increment of 17.4 – 27.5 cm in 21 days.

40 per cent paddy germinated in ExptL – I and grew from 3.1 – 24.4 cm to 14.8 – 31.5 cm, an increment of 11.7 – 7.1 cm in 9 days.

60 per cent paddy germinated in control and grew from 1.5 – 7.2 cm to 9.6 – 36.5 cm, an increment of 8.1 – 29.3 cm in 21 days.

10 per cent Malabar spinach germinated and grew in ExptL – II from 1.3 to 7.2 cm, an increment of 5.9 cm in 21 days.

10 per cent Malabar spinach germinated and grew in ExptL – I from 1.5 cm to 9.0 cm and increment of 7.5 cm in 21 days.

10 per cent Cucumber germinated and grew from 1.4 cm to 11.4 cm in ExptL –II, an increment of 10.10 cm in 21 days.

Except paddy no other seeds germinated and grew in control.

After 21 days seedlings were shifted to ground for further growth without galvanic current coverage.

During the period (between 21/7/19 to 15/8/19), the generation of galvanic current in the germination trays of ExptL – I (Al + Al) and ExptL – II (Zn + Zn) in the form of mV and µA exhibited 7 peaks and 8 valleys. From the climatic observations during the period, the peaks of 26/7/19, 2/8/19 and 5/8/19 were associated with dry and semi – dry soil of the surface in the experimental trays; while heavy rains and accumulation of water in germinating trays containing electrodes in the experimental trays caused the fall in galvanic current production representing valleys in the graph.

The production of galvanic current (mV and µA) in both the experimental troughs having similar type and similar size of electrodes (Aluminium + Aluminium and Zinc + Zinc), exhibited a similar pattern of rise and fall (peaks and valleys), but the intensity of galvanic current production in ExptL – I and ExptL – II differs significantly (Table 4, Figure 4).

A positive influence of galvanic current produced by zinc electrodes have been observed with respect to germination of sunflower, paddy, Malabar spinach and cucumber as against aluminium electrodes (40 per cent paddy and 10 per cent Malabar spinach) and control (60 per cent paddy). As regards comparative growth of paddy in zinc electrodes, the increment in growth was marginally better (17.4 – 27.5 cm) than control (8.1 -29.3 cm).

Three short time experiments (20.8.19 – 10.9.19) were taken up in plastic trays along with the control having a soil depth of 4.5 cm. The trays were of 28 X 23 cm and were provided with Aluminium + Zinc electrodes in ExptL – I and ExptL – II and with Aluminium + Aluminium in ExptL – III. The control was also run with identical conditions as that of experimental troughs, but without any electrodes. The temperature of the soil was 30°C on the day of start.

Each of the tray, both experimental and control was planted with following seeds, 1.0 cm deep under the soil

- ☆ Paddy (*Oryza sativa*) – 30
- ☆ Oats (*Avena sativa*) – 10
- ☆ Sunflower (*Helianthus annuus*) – 10
- ☆ Malabar spinach (*Basella alba*) – 10
- ☆ Green broad bean (*Vicia faba*) – 5
- ☆ Pumpkin (*Cucurbita pepo*) – 2

Galvanic current potential that was produced in ExptL – I, ExptL – II and ExptL – III ranged from 87 – 470 mV in ExptL – I (with Al + Zn electrodes); 70 – 525 mV in ExptL – II (with Al + Zn electrodes) and 11 – 257 mV in ExptL – III (with Al + Al electrodes).

Similarly the current intensity (μA) in ExptL – I, ExptL – II and ExptL – III were fluctuating between 10 μA – 150 μA in ExptL – I; 5 – 320 μA in ExptL – II and 0 – 40 μA in ExptL – III (Table 5, Figure 5).

Table 5: Measurements of Voltage and Current Intensity (At 24 hrs interval)

Date	Exp - I		Exp - II		Exp - III		Date	Exp - I		Exp - II		Exp - III	
	mV	*μA*	*mV*	*μA*	*mV*	*μA*		*mV*	*μA*	*mV*	*μA*	*mV*	*μA*
20.08.19	195	150	414	320	62	20	31.08.19	320	10	378	10	87	40
21.08.19	177	80	301	150	151	10	01.09.19	95	10	358	190	92	10
22.08.19	115	50	213	10	59	10	02.09.19	252	50	190	60	88	10
23.08.19	127	35	224	120	16	20	03.09.19	405	90	416	90	19	10
24.08.19	175	70	249	110	32	20	04.09.19	171	40	104	10	60	10
25.08.19	181	80	195	110	105	10	05.09.19	386	50	462	40	78	30
26.08.19	190	70	260	130	58	10	06.09.19	87	50	70	40	39	0
27.08.19	184	80	365	150	257	0	07.09.19	180	40	90	20	133	30
28.08.19	295	80	451	120	26	10	08.09.19	210	20	225	5	15	25
29.08.19	224	90	525	120	20	0	09.09.19	470	90	412	150	72	10
30.08.19	236	50	342	10	11	0	10.09.19	423	120	129	110	42	10

At the end of 11 days, the percentage of germination and growth of the following seedlings have been recorded.

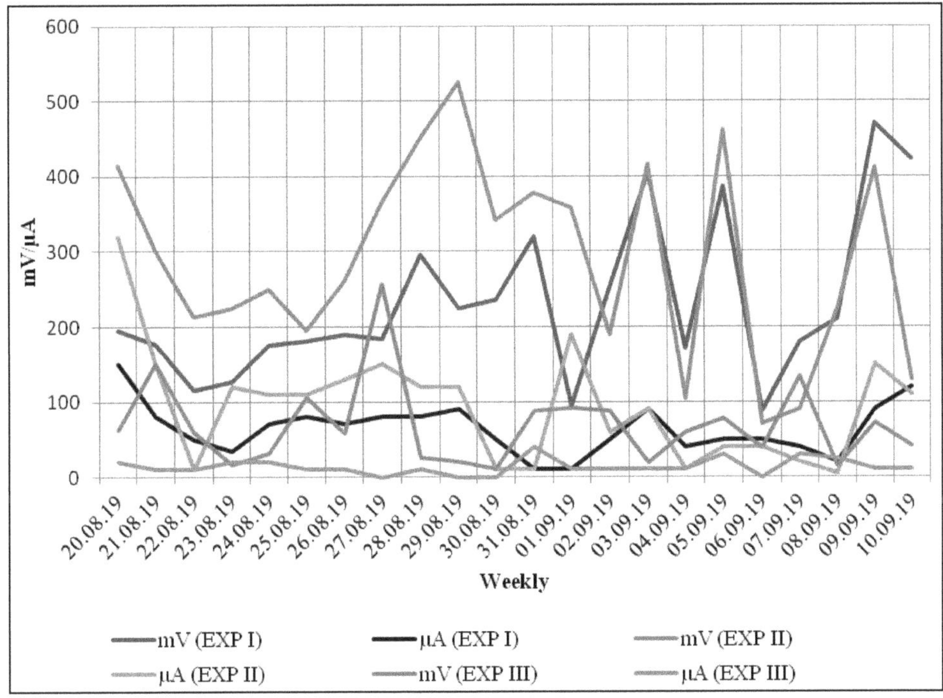

Figure 5: Pattern of Galvanic Current Intensity around Plant Seed and Seedlings.

In experiment-I, 26.7 per cent paddy germinated and grew from 0.9 – 17.2 cm to 15.5 – 24.6 cm, an increment of 7.4 – 14.6 cm within 11 days; whereas 16.7 per cent paddy germinated and grew in both Exptl –II and ExptL –III, but with different increment in growth; 6.7 – 18.3 cm (from 1.2 – 7.2 cm to 7.9 -25.5 cm) in ExptL – II and 10.7 – 22.4 cm (from 0.3 – 0.6 cm to 11 – 23 cm) in ExptL – III, which indicates a better growth in low intensity of galvanic current created by Al + Al electrodes in ExptL – III. In control, however, a low germination of 10 per cent of paddy was effected followed by medium growth rate (an increment of 4.7 – 20 cm in 11 days).

Among other seeds, only 10 per cent Malabar spinach and 10 per cent pumpkin germinated in ExptL – I and grew from 8.2 – 10.5 cm in case of the former and 8 – 11.9 cm in case of later within 11 days.

Simultaneously experiments were conducted in water media with paddy, green broad bean and Bengal gram in two glass tanks (19 X 9 cm) with a water depth of 8.5 cm, having a water temperature of 29°C and electrical resistance of 8000 Ω/cm².

Ten paddy (*Oryza sativa*), 10 Bengal gram (*Cicer arietinum*) and 2 green broad bean (*Vicia faba*) with five black molly (Ornamental fish) in one of the tank was introduced on 20.8.19. Aluminium with copper as electrodes in one tank, while no electrodes in the other tank containing fish was introduced at the beginning.

Galvanic current of 530 – 580 mV and 370 – 1700 µA was generated in the tank containing Al and Cu electrodes in most of the days except last 3 days; while 560 – 575 mV and 630 – 700 µA were produced in the tank containing fish and aluminium electrodes (Table 6, Figure 6).

Table 6: Measurements of Voltage and Current Intensity (At 24 hrs interval)

Date	*Experiment*		*Date*	*Experiment*		*Date*	*Experiment*	
	mV	*µA*		*mV*	*µA*		*mV*	*µA*
20.08.19	530	1700	25.08.19	560	370	30.08.19	580	1020
21.08.19	555	960	26.08.19	554	700	31.08.19	29	20
22.08.19	561	1210	27.08.19	575	630	01.09.19	130	0
23.08.19	567	1110	28.08.19	565	1035	02.09.19	31	10
24.08.19	540	1200	29.08.19	546	885			

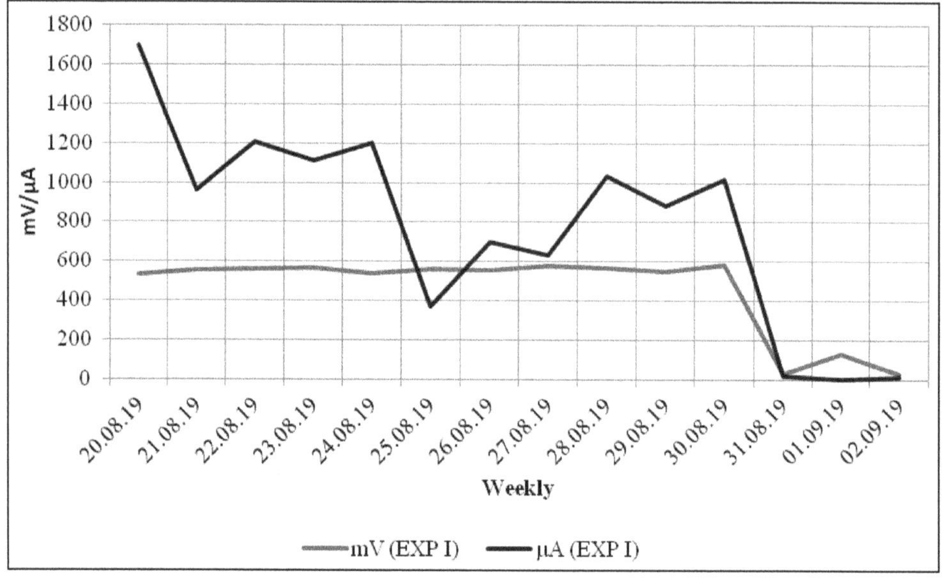

Figure 6: Pattern of Galvanic Current Intensity around Plant Seed and Seedlings.

After 6 days in the tank containing Al + Cu electrodes, 60 per cent paddy, 100 per cent Bengal gram and 100 per cent green broad bean have been germinated. In the other tank containing fish 70 per cent paddy and 100 per cent each of Bengal gram and green broad bean germinated.

In experimental tank with Aluminium + Copper electrodes, 20 per cent paddy was lost, only 40 per cent grew from 0.3 – 4.4 cm to 5.2 – 10.3 cm, an increment of 4.9 – 5.9 cm within 8 days (0.675 cm/day). Ninety percent Bengal grams grew from 2 – 10.2 cm to 9.6 – 20.2 cm an increment of 7.6 – 10 cm in 8 days (1.1 cm/day). 100 per cent green broad beans grew from 16.8 – 20 cm to 23.3 – 23.4 cm, an increment of 6.5 – 3.4 cm in 8 days (0.62 cm/day) (Figure 7).

Figure 7

In central tank, 30 per cent paddy was lost, only 40 per cent paddy grew from 0.3 – 6.3 cm to 5.6 – 14.9 cm, an increment of 5.3 – 8.6 cm in 8 days (0.87 cm/day).

30 per cent Bengal Gram grew from 9 – 9.5 cm to 14.9 – 20 cm, an increment of 5.9 – 10.5 cm in 8 days (1.025 cm/day). All the green broad beans survived in the tank.

The tank with Cu and Al electrodes could exhibit 90 per cent survival and growth of Bengal gram and 100 per cent survival and growth of green broad beans as against 30 per cent survival and growth of Bengal gram and no survival of green broad beans in tank without electrodes.

Three experiments were conducted in watery and soil media along with a control for germination of seeds with three different sets of electrodes, separated from each other by different distances. These experiments were conducted in four glass troughs with 5.1 to 8.8 cm depth of water for 7 days (31.8.19 – 6.9.19). The seeds were half immersed at the water surface on synthetic nets for soaking with water and germination. The temperature of water was 29°C and the electrical resistance at 8.3 MΩ/cm^2. The experimental set ups were;

ExptL – I – With aluminium and copper electrodes, separated by 17 cm with 5.1 cm depth of water.

ExptL – II – With aluminium and galvanized iron electrodes, separated by 37.5 cm with 5.1 cm depth of water.

ExptL – III – With copper and galvanized iron electrodes separated by 73 cm with 8.8 cm depth of water.

Control – Without any electrodes, but with 5.1 cm depth of water.

Depending on the water surface area, the number of seeds of each of the following varieties were different for each experiment and control.

	ExptL – I	*ExptL – II*	*ExptL – III*	*Control*
Corn (*Zea mays*)	5	10	20	5
Paddy (*Oryza sativa*)	25	50	100	25
Oats (*Avena sativa*)	5	10	20	5
Bengal gram (*Cicer arietinum*)	5	10	20	5
Malabar spinach (*Basella alba*)	3	6	12	3
Pumpkin (*Cucurbita pepo*)	1	2	4	1
Sunflower (*Helianthus annuus*)	5	10	20	5

At the start the galvanic current generated in each of the experimental tank was 542 mV and 480 µA in ExptL – I, 429 mV and 190 µA in ExptL – II and 996 mV and 1780 µA in ExptL – III. During 7 days in water media the galvanic current in each of the experimental tank ranged from 516 – 588 mV and 480 – 900 µA in ExptL – I; 364 – 431 mV and 190 - 470 µA in ExptL – II and 900 – 996 mV and 680 – 1780 µA in ExptL – III (Table 8, Figure 8).

Table 8: Measurements of Voltage and Current Intensity (At 24 hrs interval)

Date	8 (A)		8 (B)		8 (C)		Date	8 (A)		8 (B)		8 (C)	
	mV	*µA*	*mV*	*µA*	*mV*	*µA*		*mV*	*µA*	*mV*	*µA*	*mV*	*µA*
31.08.19	542	480	429	190	996	1780	10.09.19	633	190	321	20	1033	260
01.09.19	546	870	365	250	933	1450	11.09.19	653	190	410	40	1041	270
02.09.19	553	900	364	410	927	1200	12.09.19	680	190	460	70	1038	320
03.09.19	521	740	385	360	903	840	13.09.19	567	250	317	20	1010	70
04.09.19	516	690	381	470	927	830	14.09.19	603	160	389	70	1004	330
05.09.19	588	670	431	320	900	800	15.09.19	566	150	386	70	990	300
06.09.19	539	700	374	290	921	680	16.09.19	637	130	219	10	1000	250
07.09.19	575	240	407	100	890	250	17.09.19	504	150	445	90	1022	350
08.09.19	554	250	407	50	970	130	18.09.17	260	130	439	40	1026	330
09.09.19	604	160	327	50	1011	260							

In ExptL – I, 40 per cent of corn seed germinated after 72 hours of putting the seeds in water; while 60 per cent germination of them was noticed after 144 hours. 20 per cent, 30 per cent and 40 per cent of corn seeds germinated in ExptL – II after 72, 120 and 144 hours respectively. In ExptL – III, however 5 per cent, 20 per cent

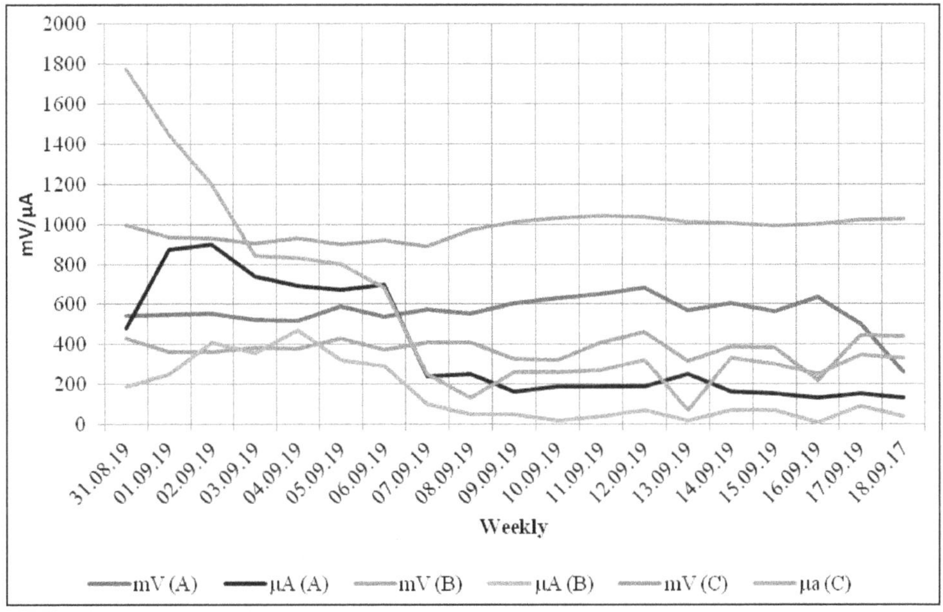

Figure 8: Pattern of Galvanic Current Production in Water Germination of Seeds and Growth of Seedlings.

and 25 per cent of corn seeds germinated after 72, 120 and 144 hours respectively. In control, only 20 per cent germination of corn seeds were noticed after 120 hours. Germination of corn seeds were more in all the experimental tanks, which indicate that there was positive influence of galvanic current on the germination of corn seeds. Highest germination (60 per cent) in ExptL – I with aluminium and copper electrodes; while the lowest germination (25 per cent) was effected in ExptL – III with copper and galvanized iron (GI) combination. The middle range of 40 per cent germination was effected with aluminium and galvanized iron combination in ExptL – II. The germination in control was even lower (20 per cent) that of the lowest (25 per cent) of galvanic induced germination of ExptL – III.

In paddy, the effect of galvanic current on germination was not very much pronounced with respect of control except 63 per cent germination in 144 hours in ExptL – III with Cu + GI electrodes combination. In 168 hours, when control exhibited 56 per cent germination, 60 per cent germination was noticed in ExptL – II with Al + GI electrodes combination. Only 44 per cent of paddy germinated in ExptL – I with Al + Cu electrodes combination. The galvanic current influence on germination of paddy may, therefore, be called as marginal.

In 72 hours, 20 per cent, 10 per cent and 5 per cent oats germinated in ExptL – I, II and III respectively. In 120 hours another 10 per cent (10 + 5 = 15 per cent) oats germinated in ExptL – III. In control 20 per cent oats germinated in 120 hours. Not much influence of galvanic current was noticed in germination of oats.

While in control 100 per cent Bengal gram germinated in 120 hours, in ExptL – I 100 per cent germination of Bengal gram was effected in 144 hours, 85 per cent germination In ExptL – III after 168 hours and only 50 per cent germination in 144 hours in ExptL – II. Galvanic current with respect of Bengal gram did not have any positive effect on germination in this case.

While 80 per cent germination was noticed within 120 hours in case of sunflower in ExptL – I, Only 40 per cent germination was observed in control within 144 hours. In ExptL – II and ExptL – III 50 per cent and 35 per cent germination of the same species took place within 120 and 144 hours respectively. Galvanic current with Al and Cu combination could effect highest germination (80 per cent) within short period of 120 hours as against 40 per cent germination in control condition within 144 hours.

60 per cent corn germinated in ExptL – I and grew from 9.8 – 14.1 cm to 12 – 33.4 cm, an increment of 2.2 – 19.3 cm within 7 days (1.54 cm/day).

40 per cent corn germinated in ExptL – II and grew from 20 cm to 29.6 cm, an increment of 9.6 cm within 7 days (1.37 cm/day).

25 per cent corn germinated in ExptL – III and grew from 4 – 9 cm to 14.6 – 24.4 cm, an increment of 10.6 – 15.4 cm within 7 days (1.86 cm/day).

None of the corn seedlings survived in control.

Therefore, the galvanic current in all the three experiments have been found to have beneficial effect, not only on the germination (higher percentage of germination) but also on the survival and enhanced growth (1.86 cm/day) in ExptL – III with GI + Cu electrodes.

56 per cent paddy germinated in control and grew from 1 – 20 cm to 10 – 30 cm, an increment of 9 – 10 cm within 7 days (1.36 cm/day).

60 per cent paddy germinated in ExptL – II and grew from 1 – 20 cm to 10 – 30 cm, an increment of 9 – 10 cm within 7 days (1.36 cm/day).

44 per cent paddy germinated in ExptL – I and grew from 1 – 15 cm to 15 – 25 cm, an increment of 10 – 14 cm within 7 days (1.71 cm/day).

63 per cent paddy germinated in ExptL – III and grew from 1 – 20 cm to 5 – 30 cm, an increment of 4 – 10 cm within 7 days (1 cm/day).

The germination rate of paddy with the influence of galvanic current was more in ExptL II and III than control. The increment in growth was found equal (ExptL - II) or more (ExptL - I) with that of control.

100 per cent Bengal gram germinated in control and grew from 14 – 15 cm to 18.2 – 20.7 cm, an increment of 4.2 – 5.7 cm within 7 days (0.71 cm/day).

100 per cent Bengal gram germinated in ExptL – I and grew from 13 – 15 cm to 17.9 – 22 cm, an increment of 4.9 – 7 cm within 7 days (0.85 cm/day).

50 per cent Bengal gram germinated in ExptL – III and grew from 11 – 13 cm to 18.2 – 18.4 cm, an increment of 5.4 – 7.2 cm within 7 days (0.9 cm/day).

Though the rate of germination of Bengal gram in control was the same of ExptL – I and even more from ExptL – III, yet the rate of growth after germination were more in experimental condition under the influence of galvanic current, from the control.

After a week, the seedlings of all the varieties were lifted from watery media and transferred to soil in the same glass tank for better nutrition and growth, with the coverage of galvanic current as before to the seedlings born and brought up under the influence of galvanic current in watery media.

The following seedlings survived after transferring them in the soil media with the influence of galvanic current.

	ExptL - I	*ExptL - II*	*ExptL - III*	*Control*
Paddy	8 (32 per cent)	19 (38 per cent)	47 (46 per cent)	10 (40 per cent)
Corn	2 (40 per cent)	1 (20 per cent)	2 (10 per cent)	Nil
Bengal gram	3 (60 per cent)	Nil	2 (10 per cent)	2 (40 per cent)

After 30 days of plantation of seedlings, the growth and vigor of the plants were examined.

With respect to paddy, the growth and vigor plants in ExptL – II was highest, followed by ExptL – I and control. Though ExptL – III had more plants yet their growth and vigor were falling behind.

As regard the corn, the growth and vigor was highest in ExptL – I followed by ExptL – III and ExptL – II.

After 12 days in the soil media with coverage of galvanic current the seedlings were transferred to the ground soil on 19.9.19 without any coverage of galvanic current and the seedlings were allowed to grow as normal plants.

After 63 days of transferring seedlings of corn to ground soil without galvanic current coverage, the corn plants grew to a length of 127 cm and 66 cm bearing two fruits (corn) in plants of ExptL – I. A single corn appeared in one plant of ExptL – II which grew to a length of 71.12 cm during that period. Highest length of 142.24 cm was attended by a single plant of ExptL – III with multiple branching bearing 4 fruits (Corns) at that time [Figures 9(a), (b), (c)]. Thus it can be said that with higher intensity of galvanic current (903 – 933 mV and 680 – 1450 µA) influence, generated by copper and galvanized iron electrodes in ExptL – III, an exceptional (branching in monocot corn stem and bearing fruits in multiple branching) a genetic modification may be possible which may be beneficial for cultivation.

9(a)

9(b)

9(c)

Figure 9a-c

Experiment was conducted in water media, in a glass tank, for 12 days along with a control with galvanized iron and copper wire lattice gate as electrodes with 10 paddy, 10 Bengal gram and 10 corn seeds in each.

The galvanic current characteristics of the experimental tank were 987 mV and 1850 µA at the start, when the temperature of water media was 28°C.

After 24 hours, the galvanic current intensity of the water media was recorded as 965 mV and 1490 µA, when 20 per cent Bengal gram seeds were found to be germinated, as against, 10 per cent of germination of Bengal gram in control. After 48 hours the galvanic current profile was 870 mV and 1710µA, showing 100 per cent germination in Bengal gram and 30 per cent germination of paddy as against 100 per cent germination of Bengal gram and 20 per cent germination in paddy in control. Galvanic current in the experimental tank have been found to fluctuate as; 840 mV and 1350 µA on 3rd day; 858 mV and 850 µA on 4th day; 835 mV and 960 µA on 5th day; 740 mV and 580 µA on 6th day; 860 mV and 1150 µA on 7th day; 860 mV and 1290 µA on 8th day; 880 mV and 740 µA on 9th day; 900 mV and 1330 µA on 10th day and 845 mV and 510 µA on 11th day (Table 10, Figure 10).

Table 10: Galvanic Current Flow through Seeds and Plant Roots

Date	*mV*	*µA*	*Date*	*mV*	*µA*
07.09.19	987	1850	13.09.19	740	580
08.09.19	965	1490	14.09.19	860	1150
09.09.19	870	1710	15.09.19	860	1290
10.09.19	840	1350	16.09.19	880	740
11.09.19	858	850	17.09.19	900	1330
12.09.19	835	960	18.09.19	845	510

After 72 hours, 50 per cent of paddy germinated in the experimental tank whereas no further germination of paddy was observed in control till the end of experiment. 20 per cent corn germinated in the experimental; tank, whereas no germination of corn was noticed in control environment.

20 per cent paddy in control grew from 2 – 3.8 cm to 7.3 – 12.6 cm, an increment of 5.3 – 8.8 cm within 5 days (1.41 cm/day). In the experimental tank, 50 per cent paddy grew from 0.8 – 4.9 cm to 6.4 – 10 cm, an increment of 5.1 – 5.6 cm within 5 days (1.07 cm/day).

Only 40 per cent Bengal gram survived both in control and experimental condition even though, the germination was 100 per cent in both the cases. 40 per cent Bengal gram in experimental condition grew from 4 – 11 cm to 9 – 16 cm, an increment of 5 cm within 5 days (1 cm/day); while in control 40 per cent of them grew from 9.5 – 13 cm to 15.7 – 19.3 cm, an increment of 6.2 – 6.3 cm within 5 days (1.25 cm/day).

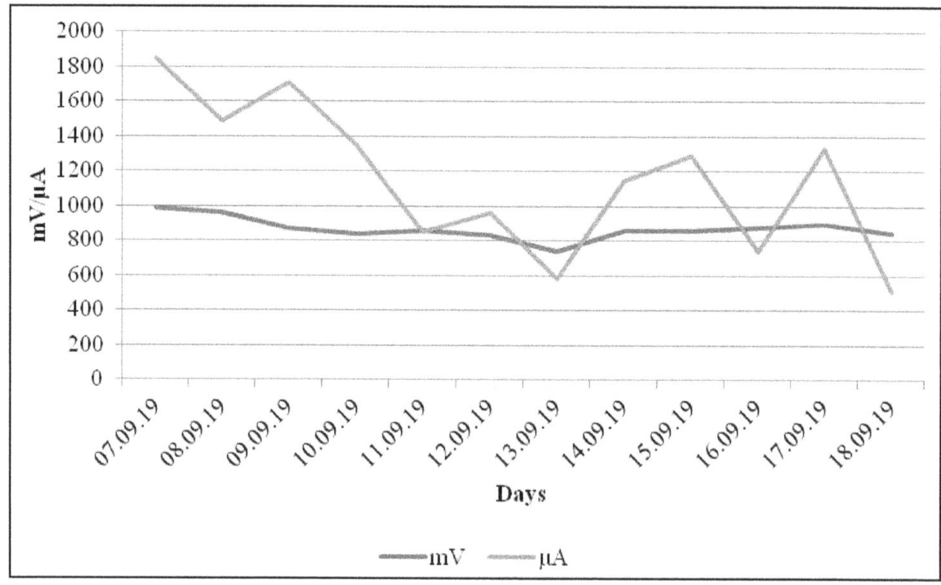

Figure 10: Pattern of Galvanic Current Production in Watery Media (Glass tank) Containing Germinating Seeds.

No corn seed germinated in control; while in experimental condition, under the influence of galvanic current 20 per cent corn germinated and grew from 4 – 4.1 cm to 20.7 – 22.8 cm, an increment of 16.7 – 18.7 cm within 5 days (3.54 cm/day).

The study reveals the sensitivity of corn seed towards galvanic current and a positive response in germination of seeds and growth of seedlings (Figures 11a and b). On the fifth day two corn plants (20 per cent of germinated seeds) were transferred to ground soil without galvanic current coverage. After 63 days in ground soil, those two corn plants attained a length of 112 cm and 95.25 cm each. The exceptional genetic modification of branching in monocot stem was noticed in smaller plant with two branches bearing one fruit in each (Figure 12) while the taller plant grew as usual bearing two fruits on the node (normal). Once again the higher intensity of galvanic current exposure, produced by galvanized iron and copper wire lattice gate electrodes in the watery media could effectively modify (genetic mutation) branching of monocot plant, bearing fruit in each branch like dicot vascular plants.

Experiments on the effect of galvanic current have been carried out for 11 days (20.9.19 to 1.10.19) on spinach (*Spinacia oleracea*), coriander (*Coriandrum sativum*), radish (*Raphanus sativus*) and methi (*Trigonella foenum graecum*) along with control. The experimental trough containing 4.5 cm soil was provided with

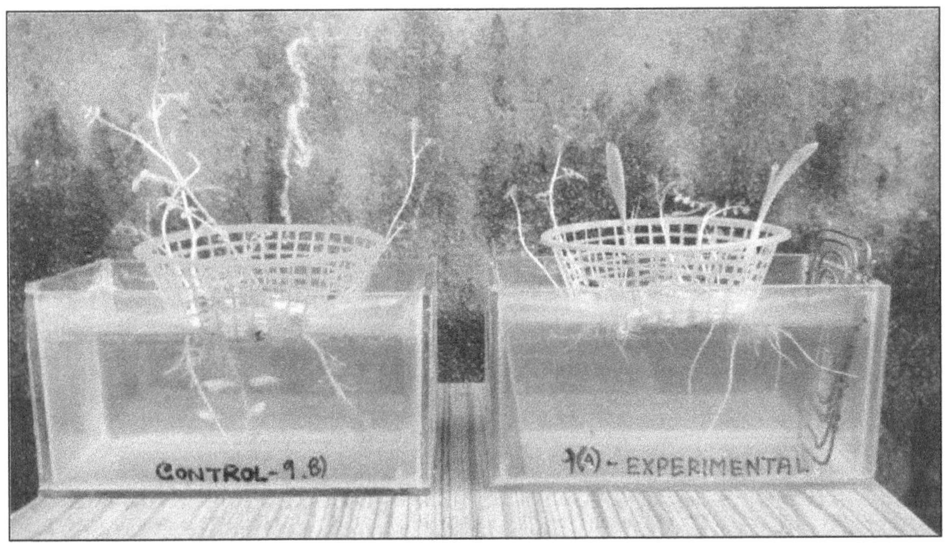

Figure 11a

Figure 11b

Figure 12

copper and aluminium plates as electrodes. Both in experimental and control trough, 100 spinach, 100 coriander, 50 radish and 50 methi seeds in each was broadcasted for germination. Coriander and spinach seed were soaked with water for 48 hours before putting them on the moist soil, while radish and methi were broadcasted on moist soil without soaking.

After 96 hours 1 methi seed out of 50 (2 per cent) germinated in the experimental tray; while during that period 2 methi seeds out of 50 (4 per cent) germinated in control. After 120 hours, one radish seed out of 50 (2 per cent) germinated in control trough.

Length measurement of methi on 1/10/19:

Experiment (cm)	Control (cm)
7.1	7, 8

The flow of galvanic current in experimental trough in between heavy rains for 7 days fluctuated, (the soil temperature being 21° - 28°C, between 25° - 27°C most of the days), from 212 to 587 mV and from 130 to 760 µA (Table 13, Figure 13).

Table 13: Galvanic Current Flow through Seeds and Plant Roots of Plants

Date	mV	µA	Date	mV	µA
20.09.19	568	550	26.09.19	520	140
21.09.19	492	620	27.09.19	516	210
22.09.19	573	760	28.09.19	587	210
23.09.19	421	170	29.09.19	449	270
24.09.19	212	130	30.09.19	451	240
25.09.19	423	320	1.10.19	460	190

Only methi seedlings survived and their growth within 8 days were 7.1 cm in experimental trough and 7 and 8 cm in control trough.

A ten days trial (2.10.19 - 12.10.19) was given in water media with 20 corn seeds (*Zea mays*) where seeds were place on net trays, so that seeds remain partially immersed in the water. Aluminium and Copper electrodes were introduced at the longest extremities, vertically to produce galvanic current through water media. A control was also run without electrodes, the other conditions being the same as that of the experimental set up. Six days later (8.10.19), 10 bengal gram (*Cicer arietinum*), 20 paddy (*Oryza sativa*), 6 sunflower (*Helianthus annuus*), 1 cucumber (*Cucumis sativus*), 1 snake gourd (*Trichosanthes cucumerina*) and 1 bitter gourd (*Momordica charantia*) were also introduced both in the experimental tank and control taking care that they remain half immersed on the net tray.

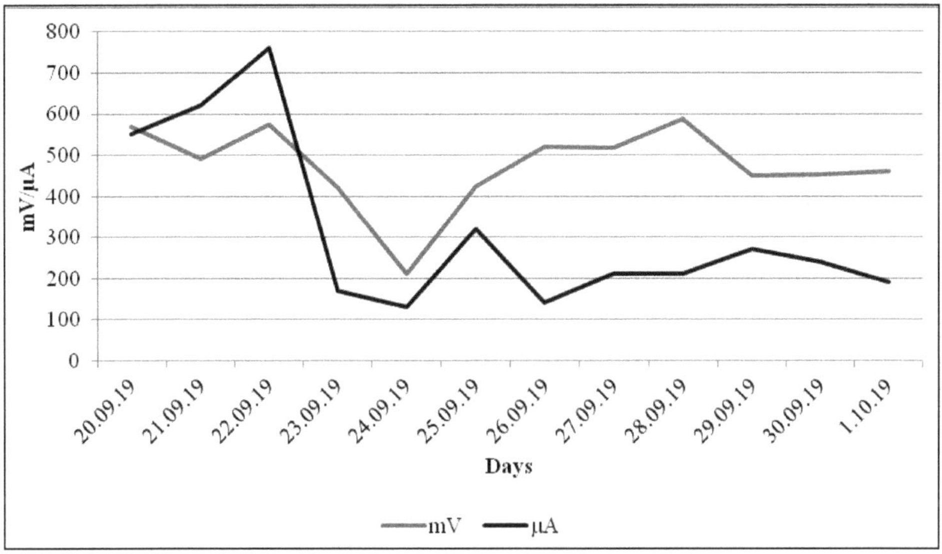

Figure 13: Pattern of Galvanic Current Flow through Seeds and Plant Roots.

The galvanic current characteristics of the experimental tank was from 395 – 550 mV and 200 – 1600 μA (Table 14, Figure 14).

Table 14: Galvanic Current Flow through Seeds and Plant Roots of Plants

Date	mV	μA	Date	mV	μA	Date	mV	μA
02.10.19	550	1600	06.10.19	530	380	10.10.19	519	200
03.10.19	486	250	07.10.19	540	510	11.10.19	523	900
04.10.19	475	430	08.10.19	395	600	12.10.19	451	250
05.10.19	535	720	09.10.19	503	390			

No germination of seeds were effected either in experimental or control even after 10 days, rather corn seeds and bengal gram have been found to be spoiled due to maggot infestations, the experiment was therefore discontinued from 13.10.19.

Three sets of experiments in soil media in the tray was taken up on 5.10.19 with Zn and Al electrodes in 15(A), Zn and Aluminium electrodes in 15(B) and Al and Copper electrodes in 15(C) along with control in identical condition except without electrodes.

Fifty paddy (*Oryza sativa*), 10 Bengal gram (*Cicer arietinum*), 10 radish (*Raphanus sativus*) and 10 methi (*Trigonella foenum - graecum*) were put in each of the experimental and control trays. The soil bed in the tray was 6 cm thick, moist and having an electrical resistance of 17 KΩ/cm². Out of the two beds with similar electrodes (Zn and Al) [Tables 15(A) and (B)], 9 volt PDC – 1/sec (external current) was applied in 15(B) from the date of starting the experiment as per details below.

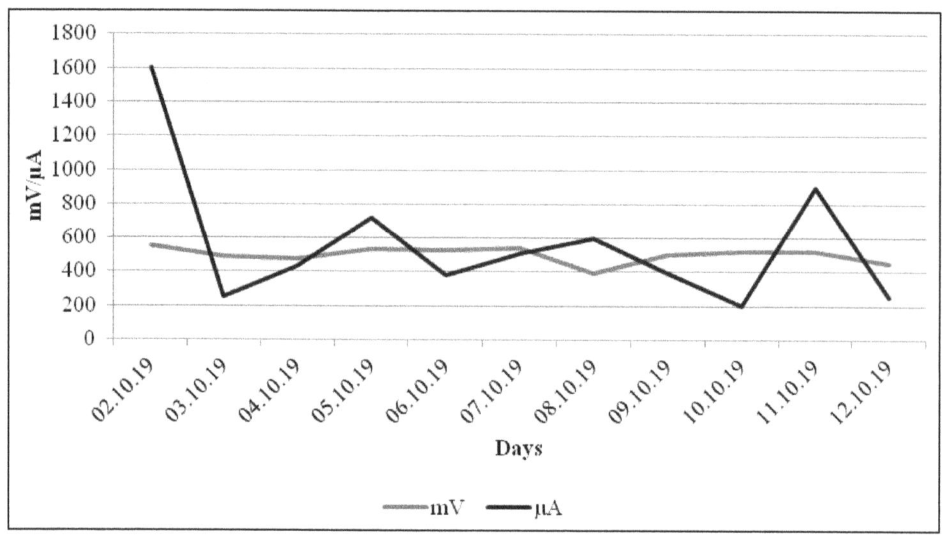

Figure 14: Pattern of Galvanic Current Flow through Seeds and Plant Roots.

Date	Duration (hours)	Applied Voltage (V)	Current Density in the Mid-field ($\mu A/cm^2$)	
			At the Beginning	At the End after Stoppage of Current
5.10.19	1	9	60	40
5.10.19	2	9	40	40
5.10.19	11	9	40	50
6.10.19	7	9	60	40
7.10.19	14	9	50	40
7.10.19	8	9	100	40
8.10.19	16	9.5	100	0
9.10.19	23	9	160	40

In all, 15(B) was subjected to external current of PDC – 1/sec, 9 volt for 82 hours in 5 days. It has been observed that in most of the days the current density in the mid – field (where seeds have been put for germination) increased during current flow, which decreased after stoppage of current flow and at times became zero.

Applying the continuous load of 1.044 KΩ in 15(C) for 6 days from 6.10.19 to 13.10.19, the current intensity increased from 90 µA to 110 -200 µA, which again dropped down to 90 µA after the removal of load. In identical electrodes (Zn and Al), the potential difference (mV) and current intensity (µA) have been found to rise from 145 mV and 100 µA to 340 – 425 mV and 110 -150 µA after application of external current in 15(B). But on stoppage of applied current the potential difference and current intensity fell down 82 – 4.8 mV and 10 – 0 µA in subsequent days. But the galvanic current of 15 (A) with Zn and Al electrodes (like that of 15 (B)) maintained a steady flow 175 – 465 mV and 50 – 160 µA with minor variation during the entire period of experiment (Table 15, Figure 15).

Table 15: Galvanic Current Flow through Seeds and Plant Roots of Plants

Date	15 (A) Galvanic current		15 (B) Applied current		15 (C) Galvanic current		Date	15 (A) Galvanic current		15 (B) Applied current		15 (C) Galvanic current	
	mV	µA	mV	µA	mV	µA		mV	µA	mV	µA	mV	µA
05.10.19	33	10	145	100	217	90	12.10.19	352	120	203	280	340	130
06.10.19	193	70	431	670	278	150	13.10.19	465	110	112	120	325	120
07.10.19	275	90	409	610	340	140	14.10.19	412	110	82	120	330	200
08.10.19	313	70	340	480	388	120	15.10.19	397	110	60	10	395	180
09.10.19	458	10	425	630	224	110	16.10.19	185	50	18.5	10	288	90
10.10.19	244	100	277	430 - 260	320	110	17.10.19	175	70	4.8	0	290	130
11.10.19	321	160	175	310 - 0	285	170							

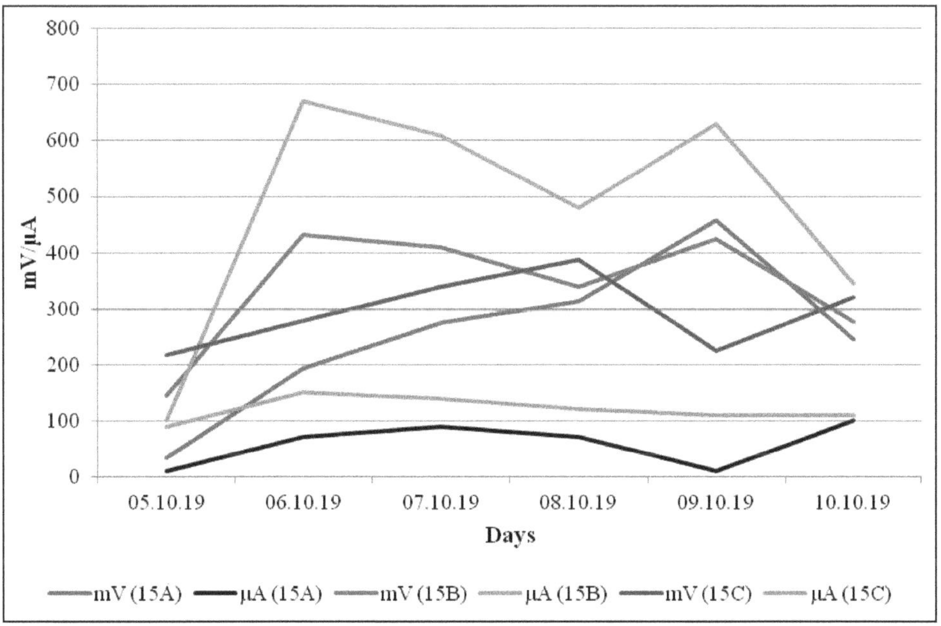

Figure 15: Pattern of Galvanic Current Flow through Seeds and Plant Roots with Three different Types of Current Exposures (15A –GC with Zn + Al, 15B – supplied DC through Zn + Al and go with extra load of 1.044 KΩ).

Only 2 seeds of radish germinated in 15 (A) covered with normal galvanic current without external intervention. No seeds germinated in 15 (B), 15 (C) and in control. Two radish seedlings, germinated in 15 (A) survived and found to grow to a length of 2.7 and 4.2 cm within 7 days.

The findings support the beneficial effect of galvanic current in germination and growth of radish seed, though in lesser rate (20 per cent).

A ten days trial for germination of seeds of *Zea mays* (Maize), *Oryza sativa* (paddy), *Helianthus annuus* (sun flower), *Raphanus sativus* (Radish) and *Cucumis sativus* (Cucumber) under the influence of galvanic current and externally applied direct current (DC) were made in soil media on trays with aluminium and copper (16A) and aluminium and aluminium (16B) as electrodes along with control.

13 maize seeds, 10 paddy, 5 sunflower, 5 radish and 1 cucumber was introduced in each experimental trays and control.

The tray with aluminium and aluminium electrodes 16(B) was subjected to externally applied DC current (9 volt) for initial 5 days for a period of 9.5 hours on the first day, 7 hours for the second day, 3 hours for the third day, 2 hours for the fourth day and 1 hour for the fifth day. The voltage at the electrode being 9V (9000 mV) which remained constant during the entire period of exposure. The density of current in the mid – field ranged 240 – 720 $\mu A/cm^2$ on the first day, 400 – 720 $\mu A/cm^2$ on the second day, 310 – 420 $\mu A/cm^2$ on the third day, 160 – 180 $\mu A/cm^2$ on the fourth day and 460 $\mu A/cm^2$ on the fifth day. After stoppage of external current, the galvanic current profile of 16 (B) was 23 – 75 mV and 0 – 30 $\mu A/cm^2$.

On the other hand the galvanic current in aluminium and copper electrode combination of 16 (A) registered an almost steady value of 442 – 625 mV and 230 – 1590 μA (Table 16, Figure 16).

Table 16: Measurements of Voltage and Current Density (At 24 hrs interval) between Electrodes Covering Seeds and Plant Roots

Date mV	16 (A) Galvanic Current		16 (B) Applied Current		Date mV	16 (A) Galvanic Current		16 (B) Applied Current	
	μA	*μA*	*mV*	*μA*		*μA*	*μA*	*mV*	*μA*
13.10.19	442	580	40	0	19.10.19	555	700	36	10
14.10.19	597	1590	615	920	20.10.19	590	1170	23	0
15.10.19	599	850	334	350	21.10.19	616	1170	36	10
16.10.19	534	430	7.2	0	22.10.19	625	1170	75	30
17.10.19	533	230	552	870					
18.10.19	567	1120	66	10					

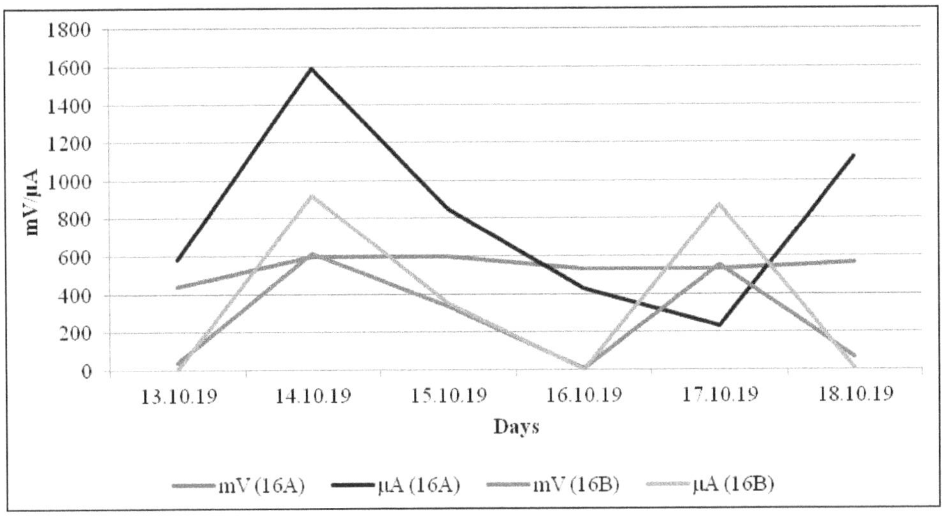

Figure 16: Pattern of Galvanic Current Flow through Seeds and Plant Roots with Two Types of Current Exposures (16A – GC with A+ + Cu Electrodes and 16B – Applied DC through Al + Al and go with extra load of 1.044 KΩ).

At this back drop, the germination of seeds; after 48 hours of putting the seeds were –

16(A) – Radish – 20 per cent

Paddy – 30 per cent

16(B) – Radish – 20 per cent

After 120 hours of sowing the seeds, the rate of germination was

16(A) – Radish – 60 per cent

Paddy – 50 per cent

16(B) – Radish – 60 per cent

Paddy – 100 per cent

It appears that the high voltage and current intensity that prevailed during the application of applied current (9000 mV and 240 – 270 μA) seems unfavorable for germination of seeds of radish and paddy. Only after stoppage of applied current after 5 day, 16(B) registered 60 per cent germination of radish and 100 per cent germination of paddy.

Comparing the results of experiments of previous set of (15(B)) where PDC 1/sec, 9volt was applied externally leading no germination of any seeds of radish and paddy with that of the later experiments of 16(B) where 9 volt continuous Dc was applied externally leading to 60 per cent germination of radish and 100 per cent germination of paddy after 120 hours, revealed that application of low voltage continuous DC has a beneficial effect on increasing the germinating rate, especially in paddy, not only from 9 volt PDC 1/sec, but also from self induced galvanic current. Thus it can be said that low voltage DC exposure in a controlled way is the most beneficial for higher rate of germination followed by self induced current than that of PDC.

Germination experiments were taken up on moist cotton (sterilized) in two plastic bowls with same copper and aluminium electrodes combination 21(A) and 21(B) along with control without electrodes. 21(A) was treated with 24 volt DC for 1 hour in two days at the beginning. On the first day from 5:20 to 6:20 pm and on the second day from 5 – 6 pm. The potential difference and intensity of current was measured with a probe of 1 cm^2 in the mid field. The electrical parameters were recorded at the time of one hour exposure and subsequently after 1, 2, 3, 4, 10 and 12 hours of stoppage of applied current. In both the days the mid field mV and µA was highest to the tune of 1393 – 1399 mV and 4207 – 4520 µA which progressively dropped down with minor fluctuation both in mV and µA and settled down at 538 – 807 mV and 350 – 470 µA.

In 21(B), the galvanic current recorded was 556 mV and 290 µA at the beginning and thereafter fluctuated from 438 – 596 mV and 70 – 490 µA prior to plantation of seedlings in ground soil. In 21(A), however, after two days of applied current at the start the galvanic current registered a variation of 404 – 761 mV and 130 – 450 µA till they were transferred to ground soil (Table 17, Figures 17A, B).

Table 17(A): Weekly Average Data of Recorded Voltage and Current Density (Tomato) between Electrodes Covering Seeds and Plant Roots

Date Range	21(A)		21(B)		Date Range	21(A)		21(B)	
	mV	*µA*	*mV*	*µA*		*mV*	*µA*	*mV*	*µA*
28.11.19 – 04.12.19	572	260	557	293	06.02.20 – 12.02.20	528	173	303	28
05.12.19 – 11.12.19	636	249	550	263	13.02.20 – 19.02.20	551	182	339	32
12.12.19 – 18.12.19	537	187	494	153	20.02.20 – 26.02.20	577	203	369	46
19.12.19 – 25.12.19	517	163	472	249	27.02.20 – 04.03.20	602	235	349	59
26.12.19 – 01.01.20	544	260	532	383	05.03.20 – 11.03.20	630	241	353	53
02.01.20 – 08.01.20	474	223	547	442	12.03.20 – 18.03.20	626	241	318	64
09.01.20 – 15.01.20	554	318	442	343	19.03.20 – 25.03.20	639	251	308	50
16.01.20 – 22.01.20	437	200	252	21	26.03.20 – 01.04.20	636	260	251	50

Date Range	21(A)		21(B)		Date Range	21(A)		21(B)	
	mV	*µA*	*mV*	*µA*		*mV*	*µA*	*mV*	*µA*
23.01.20 – 29.01.20	493	174	315	32	02.04.20 – 08.04.20	633	214	257	23
30.01.20 – 05.02.20	518	200	365	24	09.04.20 – 10.04.20	596	329	293	55

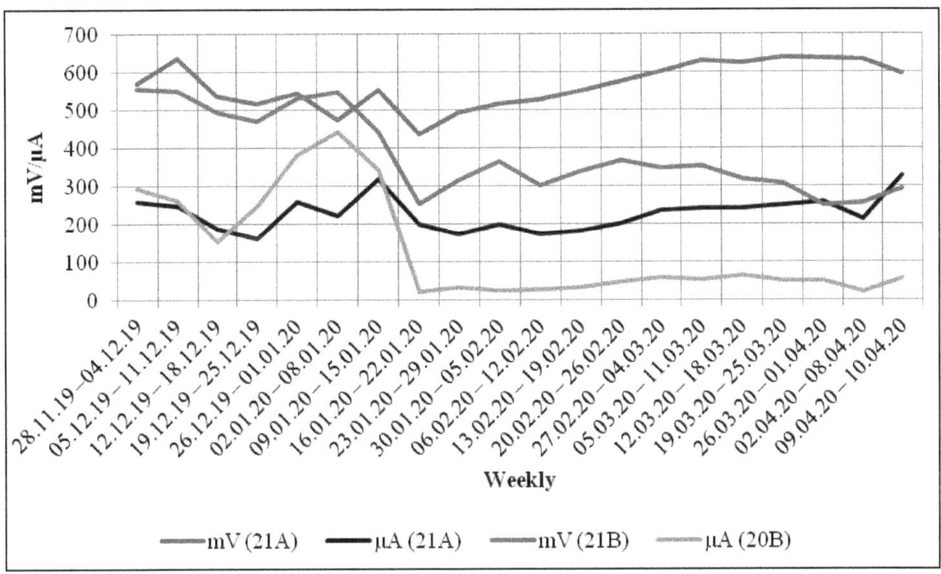

Figure 17(A): Pattern of Galvanic Current Flow through Seeds and Plant Roots of Tomato by Two Types of Current (21 A – Applied current and 21 B galvanic current).

Table 17(B): Weekly Average Data of Recorded Voltage and Current Density (Corn) between Electrodes Covering Seeds and Plant Roots

Date Range	21(A)		21(B)		Date Range	21(A)		21(B)	
	mV	*µA*	*mV*	*µA*		*mV*	*µA*	*mV*	*µA*
01.12.19 –07.12.19	602	214	559	300	23.02.20 – 29.02.20	650	327	573	193
08.12.19 – 14.12.19	866	830	546	140	01.03.20 – 07.03.20	582	335	559	207
15.12.19 – 21.12.19	735	561	474	127	08.03.20 – 14.03.20	841	566	588	199
22.12.19 – 28.12.19	690	460	463	126	15.03.20 – 21.03.20	813	448	606	246
29.12.19 – 04.01.20	644	333	506	152	22.03.20 – 28.03.20	855	624	619	257

Date Range	21(A)		21(B)		Date Range	21(A)		21(B)	
	mV	*μA*	*mV*	*μA*		*mV*	*μA*	*mV*	*μA*
05.01.20 – 11.01.20	745	238	518	154	29.03.20 – 04.04.20	523	13	673	308
12.01.20 – 18.01.20	691	423	567	228	05.04.20 – 11.04.20	653	124	640	181
19.01.20 – 25.01.20	804	425	591	222	12.04.20 – 18.04.20	612	249	716	297
26.01.20 – 01.02.20	765	401	548	190	19.04.20 – 25.04.20	556	61	645	284
02.02.20 – 08.02.20	694	366	594	239	26.04.20 – 02.05.20	542	106	579	190
09.02.20 – 15.02.20	764	500	579	222	03.05.20 – 06.05.20	400	0	618	275
16.02.20 – 22.02.20	714	717	566	190					

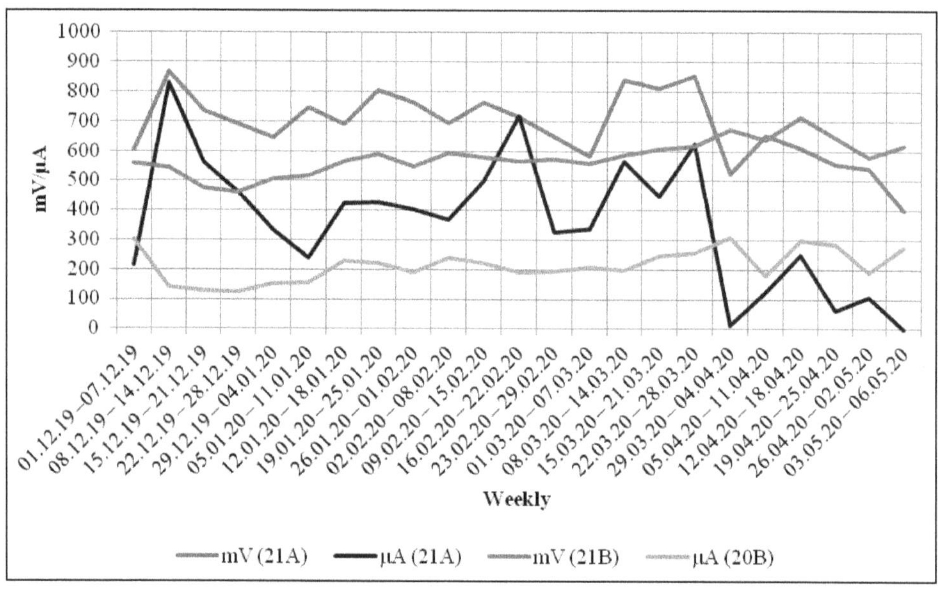

Figure 17(B): Pattern of Galvanic Current Flow through Seeds and Plant Roots of Corn by Two Types of Current (21 A – Applied current and 21 B galvanic current).

Each of the experimental bowl and control bowl, 8 tomatoes (*Solanum lycopersicum*), 2 pumpkin (*Cucurbita pepo*) and 8 corn (*Zea mays*) were put for germination.

After 48 hours - In 21(A) – 25 per cent tomato seeds germinated

37.5 per cent corn seeds germinated

In 21(B) – 50 per cent tomato seeds germinated

62.5 per cent corn seeds germinated

In control – 37.5 per cent corn seeds germinated

After 72 hours - In 21(A) – 75 per cent tomato seeds germinated

In 21(B) – 87.5 per cent tomato seeds germinated

In control – 37.5 per cent tomato seeds germinated

After 96 hours - In 21(A) – 100 per cent tomato seeds germinated

In 21(B) – 87.5 per cent tomato seeds germinated

In control – 75 per cent tomato and corn seeds germinated

Higher percentage of germination of tomato and corn within 48 hours have been noticed (50 per cent and 62.5 per cent) in galvanic current influenced bowl than with the influence of applied current and only after 96 hours 100 per cent germination in tomato was noticed in applied current and in corn only 37.5 per cent germination was noticed. The influence of galvanic current was found to the beneficial for early germination of higher percentage of tomato and corn seeds (Figure 18).

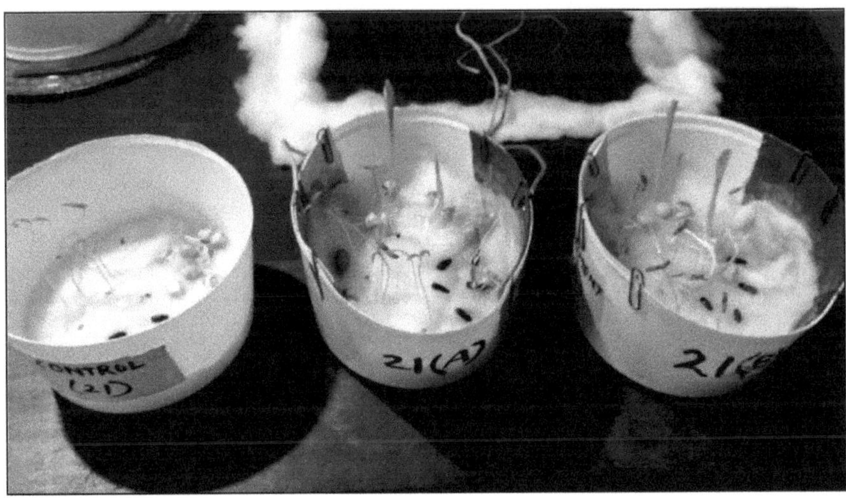

Figure 18

After 7 days of germination, tomato seedlings attained a length with the range of 4.3 – 6.2 cm in 21(A), 4.8 – 6.0 cm in 21(B) and 3.5 – 5 cm in control. The corn seedlings, however, during the period have grown to 3 – 7.5 cm in 21(A), 3 – 7 cm in 21(B) and 1.5 cm in control. After 35 days of germination, tomato seedlings attained a length of 5.5 – 9.8 cm in 21(A), 5 – 8.2 cm in 21(B) and 6.5 – 7.6 cm in control. Corn seedlings during the period attained a length of 15.6 – 25.5 cm in 21(A), 20.1 – 26.8 cm in 21(B) and 18.3 – 38.3 cm in control. During this period the seedlings of tomato and corn were transferred into ground soil, four of them, with the coverage of galvanic current by the same electrode and others without electrodes around them. The galvanic current around the tomato plant after transferring in the ground soil, ranged between 399 – 682 mV and 117 – 395 µA in 21(A), and 83 – 398 mV and 10 – 80 µA in 21(B). As regards the corn seedlings in ground soil, the galvanic current coverage was 366 – 1039 mV and 110 – 1140 µA in 21(A) and 403 – 727 mV and 70 – 400 µA in 21(B) (Figure 19). Except one tomato in one plant of 21(B), no other plants of 21(A), 21(B) and control bore fruit even though they grew to 1.2 – 1.7 m with 3 – 5 branches having biomass of 50 – 105 g each in 131 days. The plants of 21(B) were 1.3 – 1.4 m long with 3 – 4 branches in each plant and a biomass of 150 -200 g each during the same period of 131 days, after which all of them died. Corn plants were growing very tall, 1.5 – 2.0 m in length and 3 plants of 21(B) have bourne three maizes, but broken during cyclone on 138th day (Figure 20).

Fig. 19

Here also, 24 volt continuous DC exposure for 2 hours in the beginning was found to have effect on higher percentage of germination of tomato seed in the long run.

A 38 days trial was given in 4.5 cm soil bed on plastic trays with three electrode combinations, namely, Zn + Al in Expt – I, Zn + Al in Expt – II and Cu + Al in Expt – III, besides control in identical conditions without electrodes. Of these same electrode combinations (Zn + Al) in Expt – I and Expt – II, 9 volt DC was applied after 24 hours in Expt – I for four consecutive days from 23.10.19 to 26.10.19 for 1 hour on the first day (23.10.19), 2 hours on the second day (24.10.19), 3 hours on the third day (25.10.19) and 3.5 hours on the fourth day (26.10.19) at an interval of 24 hours in between. The density of current at the beginning and after one hour of stoppage of applied current were, 190 µA and 120 µA on the first day; 100 µA and 80 µA on the second day; 60 µA and 50 µA on the third day and 180 µA and 40 µA on the fourth day.

Fig. 20

No external current was applied to Expt – II and Expt – III and planted seeds were under the influence of galvanic current only produced by the inbuilt electrodes.

Each of the three experimental trays, provided with 4.5 cm moist soil, on which the seeds have been planted, bore an electrical resistance of 8.2 KΩ/cm^2 having a temperature of 28°C. Seeds put on each of the tray including control were green broad bean (*Vicia faba*), gourd (*Lagenaria sinceraria*), pumpkin (*Cucurbita pepo*) and tomato (*Solanum lycopersicum*) at the rate of 6, 3, 5, 8 numbers respectively in all the three experimental trays, except in the control where the rate was in the order of 5, 3, 5, 7 respectively.

The galvanic current characteristics in all the three experimental trays were 475 – 101 mV and 470 – 10 µA in Expt – I; 657 – 50 mV and 190 – 20 µA in Expt – II and 520 – 300 mV and 320 – 70 µA in Expt – III during the entire period with peaks and vallies. The average weekly voltage and current density of the three experimental troughs were; 178 – 334 mV and 30 – 132 µA in Expt – I; 167 – 355 mV and 34 – 83 µA in Expt – II and 349 – 445 mV and 91 – 180 µA in Expt – III (Table 21, Figure 21). The higher current parameters in Expt – III was due to combination of copper and aluminum electrodes.

Table 21: Weekly Average Data of Recorded Voltage and Current Density between Electrodes Covering Seeds and Plant Roots

Date Range	Exp - I		Exp - II		Exp - III	
	mV	*μA*	*mV*	*μA*	*mV*	*μA*
22.10.19 – 28.10.19	285	132	355	83.3	445	180
29.10.19 – 04.11.19	334	72	339	65	414	163
05.11.19 – 11.11.19	309	39	281	41	349	91
12.11.19 – 18.11.19	232	30	226	33	356	97
19.11.19 – 25.11.19	203	46	167	29	425	126
26.11.19 – 30.11.19	178	34	239	34	436	144

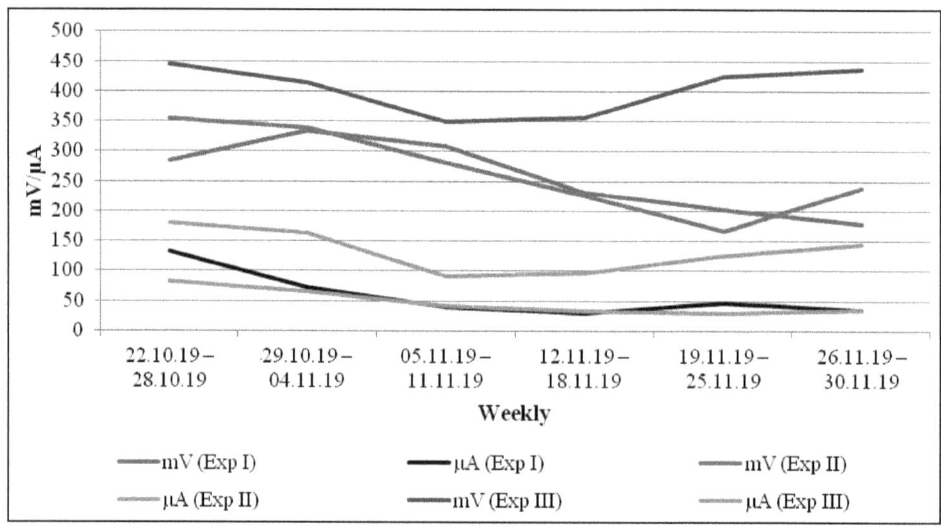

Figure 21: Pattern of Galvanic Current Flow through Seeds and Plant Roots with Three Electrode Combinations (Exp – I with Zn + Al, Exp – II with Zn + Al and Exp – III with Cu + Al)

Within 48 to 72 hours of galvanic current influence two tomatoes in each of the three experimental trays germinated and grew to 4.5 – 7.0 cm in Expt – I within 35 days; 9.2 – 10.5 cm in Expt – II within 35 days and 6.4 – 6.5 cm in Expt – III within 28 days. The tomato seedlings were finally transferred to the ground soil after 38 days from the experimental trays, where they grew to 12 cm, 21.5 cm and 13 – 15.5 cm belonging to Expt – I, Expt – II and Expt – III respectively. No tomato seed germinated in control. Only one pumpkin in Expt – I and in control germinated after 72 hours. But the seedling of Expt – I was eaten away by rat, while that of control grew upto 21 cm before transferring it to ground soil.

Fifty seeds of ladies finger (*Abelmoschus esculentus*) and 4 Chichinga (*Trichosanthes cucumerina*) seeds each in experimental pot and control were planted on 28.1.20 in soil media. In the experimental pot aluminium and copper electrodes were provided at a distance of 11 cm. In between the electrodes, seeds were planted below 1 cm of the soil surface. After 2 days, 9 seeds of ladies finger out of 50 (18 per cent) germinated in the experimental pot; while in control 7 seeds out of 50 (14 per cent) after 48 hours of sowing. After 264 hours total germination of ladies finger seeds in the experimental pot became 11 out of 50 (22 per cent) and all the 11 seedlings of the experimental pot were found healthy and growing. In control, though 13 seeds out of 50 (26 per cent) germinated, but they were weak, curling, short and only 2 survived.

After 47 days, the seedlings were transferred to the ground soil in separate plots of same size for experimental and control. In the experimental plot, the seedlings were planted between aluminium and copper plates which were inserted in the ground soil at a depth of 4 cm and separated at a distance of 40 cm from each other.

At the time of plantation of the seedlings, their lengths were 4.5 – 10.5 cm (11 seedlings) in experimental plot, while there were 10 seedlings (2.6 – 10 cm) in control plot of which 8 seedlings died and only two seedlings survived (5 – 5.3 cm).

While in the pot, the galvanic current generation around seedlings were 182 – 604 mV and 190 – 470 µA. But after transplantation of the seedlings from the pot to ground soil, fortified with galvanic current coverage, their values ranged from 523 – 731 mV and 40 – 510 µA for remaining 53 days (Table 22, Figure 22).

Table 22: Weekly Average Data of Recorded Voltage and Current Density between Electrodes (Al + Cu) Covering Seeds and Plant Roots

Date Range	*mV*	*µA*	*Date Range*	*mV*	*µA*
28.01.20 – 03.02.20	557	354	17.03.20 – 23.03.20	693	406
04.02.20 – 10.02.20	560	287	24.03.20 – 30.03.20	688	405
11.02.20 – 17.02.20	528	211	31.03.20 – 06.04.20	700	435
18.02.20 – 24.02.20	580	167	07.04.20 – 13.04.20	568	424
25.02.20 – 02.03.20	640	246	14.04.20 – 20.04.20	639	390
03.03.20 – 09.03.20	683	286	21.04.20 – 27.04.20	629	309
10.03.20 – 16.03.20	642	327	28.04.20 – 06.05.20	212	232

During this period the length of the plants in experimental plot ranged from 18 cm to 87 cm. 50 per cent of them started flowering and fruiting at the rate of 3 – 4 flowers and fruits per plant. Two plants in control plot did not survive for more than 8 days of transplantation.

No manures either organic or inorganic and pesticides were used in any of the experimental and control plots.

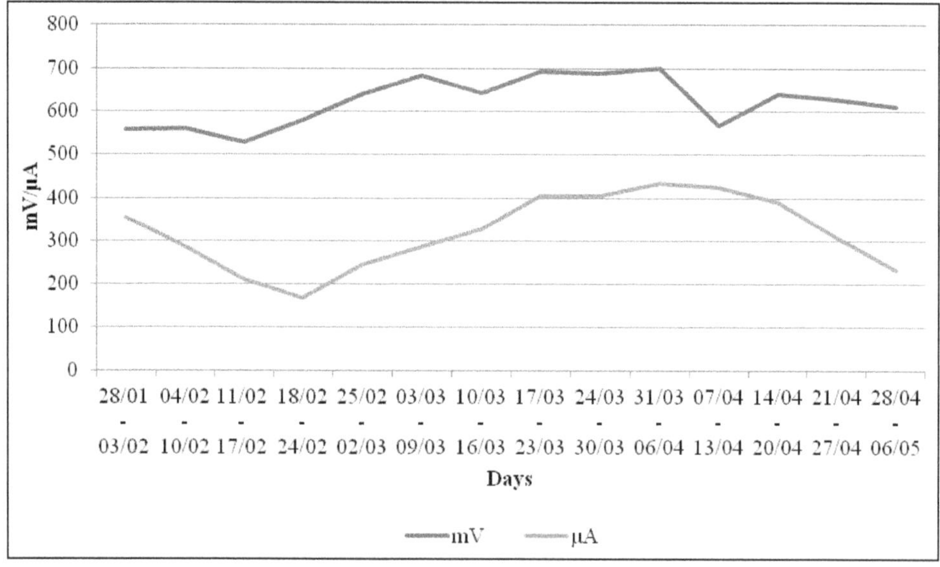

Figure 22: Pattern of Galvanic Current Flow through Seeds and Plant Roots with Al and Cu Electrodes.

Experiment was conducted between 11.9.19 to 22.9.19 along with control for germination of paddy (*Oryza sativa*), corn (*Zea mays*) and Bengal gram (*Cicer arietinum*) to observe the influence of induced galvanic current on germination and subsequent growth of seedlings. The seeds (Paddy – 1000), (corn – 30) and (Bengal gram – 30) were put under moist soil in plastic troughs containing 4.5 cm thick soil. The initial soil temperature was 27.5° C and the temperature during the test period (11 days) varied between 26° C to 30° C both in experimental and control troughs. Aluminium and galvanized iron electrodes (plates) were put in the experimental trough to produce galvanic current.

At the start, (11.9.19), the galvanic current registered a potential difference of 500 mV and current intensity of 310 µA around the seeds, which fluctuated from 342 mV to 500 mV and from 150 to 310 µA during the test period (Table 23, Figure 23).

Table 23: Galvanic Current Flow through Seeds and Plant Roots

Date	*mV*	*µA*	*Date*	*mV*	*µA*	*Date*	*mV*	*µA*
11.09.19	500	310	15.09.19	395	210	19.09.19	440	180
12.09.19	467	210	16.09.19	405	230	20.09.19	429	250
13.09.19	397	150	17.09.19	431	190	21.09.19	406	180
14.09.19	405	170	18.09.19	395	210	22.09.19	342	170

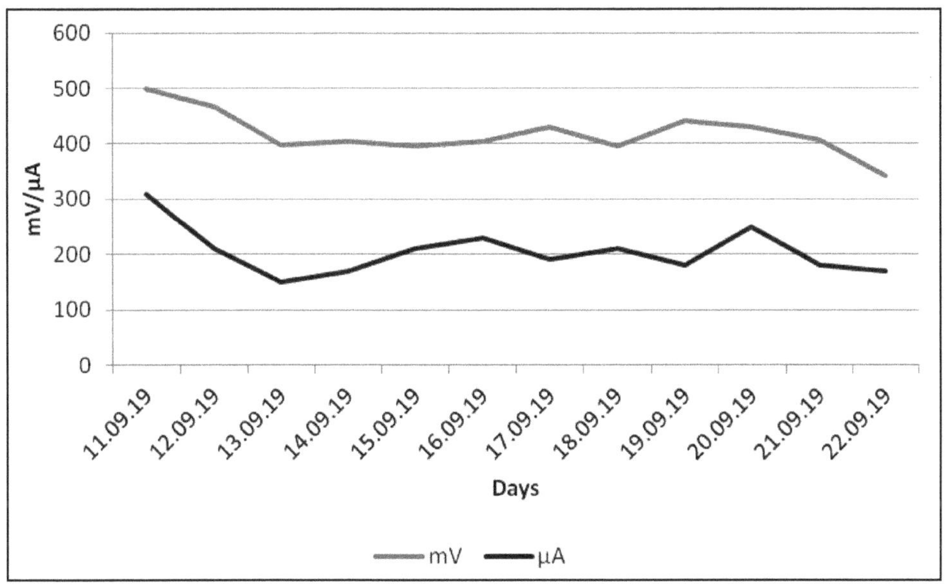

Figure 23: Pattern of Galvanic Current Flow through Seeds and Plant Roots.

After 72 hours of sowing seeds, 4 Bengal gram out of 30 (13.33 per cent) germinated in experimental trough, while 10 per cent of them (3 out of 30) germinated in control trough. After another 24 hours, only 1 seed of paddy out of 1000 seeds germinated in control. Subsequently on the 8th day 3 paddy seeds germinated in experimental tray. None of the corn seeds germinated either in experimental and control trough. The corn seeds both in experimental and control trough have been found to be infested with maggot after soaking with water.

Within 10 days Bengal gram grew to 14.4 – 19.0 cm in experimental trough, while in control their growth was recorded 9.6 – 17.5 cm, an increment of 1.44 – 1.9 cm/day under the influence of galvanic current as against 0.96 – 1.75 cm per day in control.

Three paddy (0.03 per cent) in experimental trough grew from 5 – 15.6 cm in 13 days as against a single paddy in control which grew to 8.6 cm.

All Bengal gram seedlings with in experimental and control troughs have been found destroyed due to heavy rainfall on 22.9.19.

Though, not statistically established due to insufficient data, yet the positive influence of galvanic current on germination and initial growth have been noticed in case of Bengal gram and paddy.

Ten cucumber (*Cucumis sativus*), 10 luffa (*Luffa acutangula*) and 10 ladies finger (*Abelmoschus esculentus*) seeds were subjected to the influence of galvanic current in soil media in the pot for 70 days with the help of copper and aluminium plate electrodes along with control in identical conditions with same number and types of seed but without electrodes in control for germination and growth. After 4 days of putting seeds into the soil, 5 cucumbers (50 per cent) and 2 luffas (20 per cent) germinated in experimental pot provided with copper and aluminium electrodes to produce induced galvanic current around the seeds. While in control (40 per cent) cucumber seeds sprouted. After 19 days 5 seedlings of cucumbers, and one seedling of luffa had grown to 6 – 7.5 cm and 8.5 cm respectively in experimental pot; while in control only 2 cucumber seedlings registered a growth of 9 – 10.2 cm. Two ladies finger in each, experimental and control pots were 5.5 – 8 cm and 4.5 – 6 cm in length respectively. After 22 days, 8 (80 per cent) seedlings of ladies finger in experimental pots were found to grow from 4 to 7.5 cm; while 5 (50 per cent) seedlings of them had grown to 3 to 9.2 cm in length in control pot.

During this period the galvanic current potentials and current intensities varied from 472 – 772 mV and 160 – 820 µA, the weekly average being 601 – 680 mV and 296 – 476 µA (Table 24, Figure 24).

Table 24: Weekly Average Data of Recorded Voltage and Current Density between Electrodes (Al + Cu) Covering Seeds and Plant Roots

Date Range	mV	µA	Date Range	mV	µA
27.02.20 – 04.03.20	664	403	02.04.20 – 08.04.20	676	451
05.03.20 – 11.03.20	639	381	09.04.20 – 15.04.20	648	411
12.03.20 – 18.03.20	665	476	16.04.20 – 22.04.20	633	399
19.03.20 – 25.03.20	631	349	23.04.20 - 29.04.20	601	319
26.03.20 – 01.04.20	680	472	30.04.20 – 06.05.20	626	296

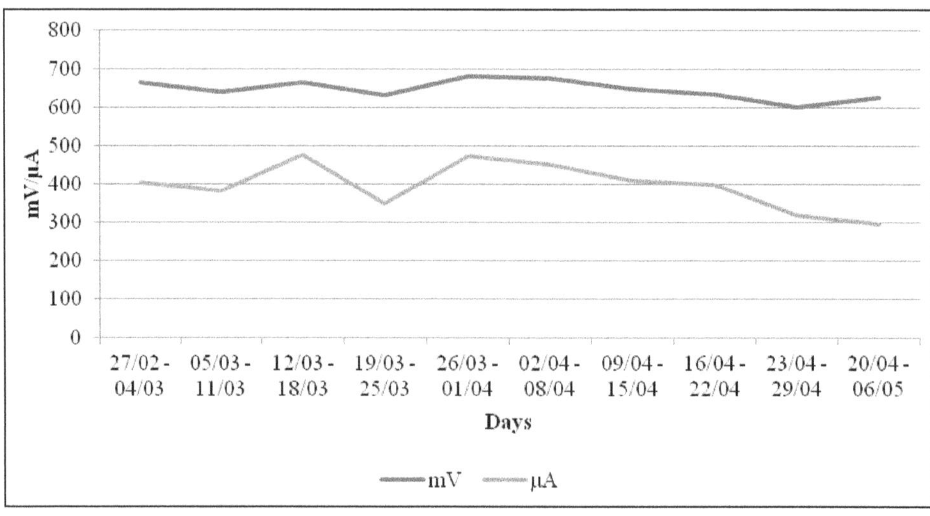

Figure 24: Pattern of Galvanic Current Flow through Seeds and Plant Roots with Cu and Al Electrodes.

80 per cent germination of ladies finger, 50 per cent germination of cucumber and 10 per cent germination of luffa in galvanic current influenced media (soil) were found beneficial than 50 per cent germination of ladies finger and 20 per cent germination of cucumber in control without galvanic current influence. Luffa did not germinate in control.

A 17 days germination program was taken up to study the influence of induced galvanic current on the seeds of paddy (*Oryza sativa*), Bengal gram (*Cicer arietinum*), Corn (*Zea mays*), sunflower (*Helianthus annuus*) and Cucumber (*Cucumis sativus*) from 14.9.19 to 1.10.19 in soil media. Four cm garden soil was used in plastic troughs with the following electrode combinations in three experimental troughs. Expt – I was provided with Zn and GI electrodes; while in Exptl – II and Exptl – III Zn and Al and Al + Cu were used respectively. A control was run in identical conditions with that of experimental ones, except without electrodes. The different combinations of electrodes in different experiments (Expt – I, II and III) induced the galvanic currents of different magnitudes depending on the nature of electrodes.

The following numbers of seeds of the above mentioned species were put in each of the experimental and control troughs.

- ☆ Paddy – 50 nos. in each of the experimental trough including control.
- ☆ Bengal gram – 10 nos. in each of the experimental trough including control.
- ☆ Corn – 10 nos. in each of the experimental trough including control.
- ☆ Sunflower – 10 nos. in each of the experimental trough including control.
- ☆ Cucmber – 2 nos. in each of the experimental trough including control.

Cloudy and sunny weathers with moderate to heavy rains prevailed during the period of test, and the soil temperature fluctuated between 20° to 29.5°C. The galvanic current characteristics among the three experimental troughs, the potential difference (mV) and current intensity (µA) were the lowest (0.3 – 167 mV/10 – 110 µA) in Expt – I, provided with Zinc and galvanized iron electrodes. The next higher potential difference (155 – 460 mV) and current intensities (40 – 140 µA) was noticed in Expt – II provided with Zinc and Aluminium electrodes. Highest potential difference and current intensities of the series (250 – 427 mV and 20 – 290 µA) were recorded in Expt – III with aluminium and copper electrode combinations (Table 25; Figure 25).

Within 48 hours 20 per cent and 40 per cent Bengal gram germinated in Expt - I and Expt – III respectively; while in control only 30 per cent Bengal gram germinated. Only 1 cucumber seed out of 2 germinated in control. Only two corn out of 10 germinated in Expt – I. One paddy out of 50 seeds germinated in control only.Sunflower seeds did not germinate in any trays (Expt or control).

During the period of culture (17 days), Bengal gram seedlings grew to a length of 14.8 – 16 cm in Expt – I while in control one seedling of Bengal gram attained a length of 19.2 cm. In Expt – I out of 2 corn seedlings, one eaten by the pest at 4.8 cm length, while the other one grew to a length of 19.9 cm during the culture period.

Table 25: Measurements of Voltage and Current Density (At 24 hrs interval) between Electrodes Covering Seeds and Plant Roots

Date	I		II		III		Date	I		II		III	
	mV	*μA*	*mV*	*μA*	*mV*	*μA*		*mV*	*μA*	*mV*	*μA*	*mV*	*μA*
14.09.19	33	20	335	140	321	170	23.09.19	9	10	412	70	268	90
15.09.19	81	70	304	50	427	250	24.09.19	0.3	25	155	50	380	290
16.09.19	91	50	359	80	256	190	25.09.19	7.6	40	284	90	390	40
17.09.19	98	70	332	80	299	210	26.09.19	8.7	10	280	90	417	20
18.09.19	55.70	40	390	80	362	280	27.09.19	13.3	20	348	60	320	160
19.09.19	57.01	50	364	50	296	190	28.09.19	25.3	10	310	40	415	190
20.09.19	60	60	372	70	250	150	29.09.19	36	30	363	70	300	110
21.09.19	17.30	10	330	80	315	200	30.09.19	62	40	420	90	345	150
22.09.19	20	20	305	70	295	180	01.10.19	167	110	460	100	342	130

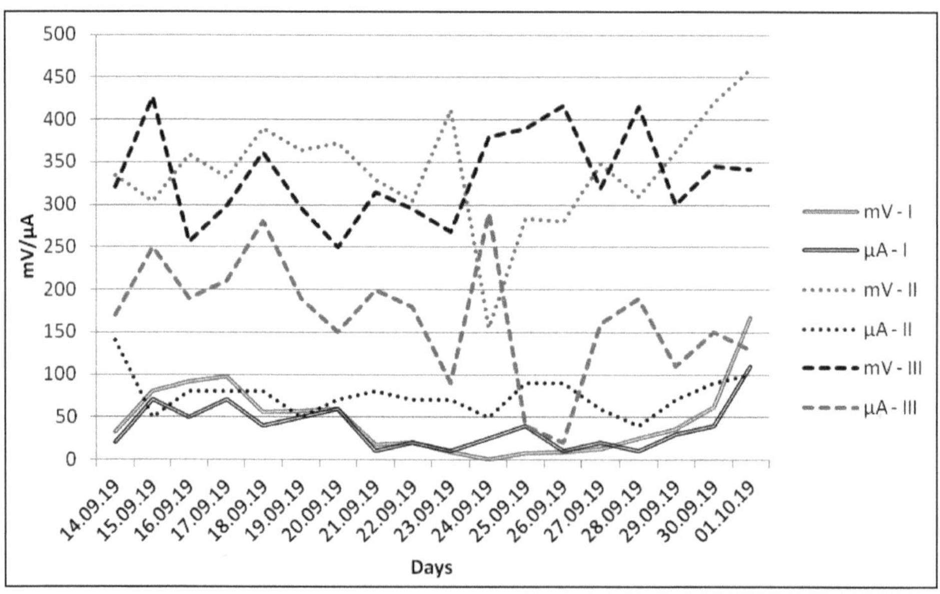

Figure 25: Pattern of Galvanic Current Flow through Seeds and Plant Roots with Three different Electrode Combinations I (Zn + GI), II (Zn + Al) and III (Al + Cu).

In the control on the other hand one Bengal gram seedling have grown to 19.2 cm after 17 days, followed by 1 paddy (22.0 cm) and 1 cucumber (13.5 cm).

It has been observed that the paddy did not germinate under the influence of galvanic current (only 1 paddy germinated and grown in control), whereas, corn did not germinate in control without influence of galvanic current (2 corn seeds germinated in Expt – I but one survived and grown) during culture period. The germaination and survival of Bengal gram was 30 per cent in control, but only one survived and grown to 20 cm. One each in Expt – II and III though germinated to a length of 2 cm and 1.5 cm respectively, yet they did not survived due to heavy rainfall and accumulation of water. In Expt – I though two Bengal gram seed germinated and grown upto 12.5 – 18 cm within 7 days, but could not survive due to heavy rainfall and water accumulation.

9

Experiments

Experiments 19(A) and 19(B) along with the control started from 29/10/19 in soil beds on glass tank with copper and aluminium electrodes in 19(A) and with only aluminium electrodes in 19(B). The distance between Al and Cu electrodes in 19(A) was 8 cm; while that of AL +AL electrodes in 19(B) was 19 cm. Seeds of tomato (*Solanum lycopersicum*) were placed centrally in between the electrodes in a row to a depth of 1 cm from the soil surface. 14 seeds in 19(A), 13 seeds in 19(B) and 13 seeds in control were planted. The initial induction of galvanic current after placing the electrodes in the soil around planted seeds were 475 mV and 500 µA in 19(A) and 30 mV and 0 µA in 19 (B) respectively.

2 seeds out of 14 (14.3 per cent) germinated after 72 hours in 19 (A) and thereafter 1 seed after 96 hours and another 1 seed after 144 hours of putting them under the soil, thus totaling 4 seeds out of 14 (28.6 per cent) in 19 (A). 1 seed out of 13 in control germinated after 216 hours.

Due to production of small amount of galvanic current in 19 (B) (30 mV and 0 µA) external current of 9 volt of PDC 1/sec was applied for three consecutive days for a period of 30 minutes on the firsts day and thereafter for 60 minutes and 90 minutes on second and third days after an interval of 24 hours between the external current application. The residual effect on galvanic current was checked by measuring the mV and µA, immediately 5 minutes after stoppage of applied current and then after 1, 2, 3, 4, 5 and 13 hours. On the first day the potential difference and current intensity rose to as high as 690 mV and 320 µA and then declined to 30 mV and 0 µA within 4 hours. With 60 minutes of current application on the second day, from 627 mV and 400 µA the galvanic current was reduced to 23 mV and 0 µA within 5 hours. When the external current was applied for 90 minutes on the third day, 499 mV and 200 µA, recorded after 1 hour of stoppage of applied current was decreased to 8.4 mV and 0 µA after 13 hours. It was observed that externally applied current may increase the galvanic current intensities in the beginning for few hours temporarily, but finally fell down to its initial level or even less (as low as 8.4 – 9 mV and 0 µA after 13 hours or more) and thereafter, may not have any effect on germination of seeds and growth of seedlings (Table of Applied Current).

Replacing one of the aluminium electrodes with copper in 19 (B) after five days, the initial value of galvanic current rose to 453 mV and 190 µA immediately after replacement. Eleven days after replacing with copper electrode, one of the tomato seed sprouted in 19 (B). There within a week two more tomato seeds germinated in 19 (B).

It was therefore evident, that for germination of tomato seeds, some minimum values of galvanic current intensities were required, which AL +AL electrodes in 19 (B) failed to produce till one of the Al electrodes was replaced by copper electrode. The galvanic current intensities of 19 (A) and 19 (B) within 133 days fluctuated 399 – 784 mV and 140 – 820 µA in 19 (A) and from 435 – 702 mV and 70 – 750 µA in 19 (B) after changing the aluminium electrode with copper electrode. The weekly average of mV and µA during the period ranged from 474 – 687 mV and 321 – 766 µA in 19 (A) and 161 – 743 mV and 62 – 409 µA in 19 (B) (Table 26, Figure 26).

Table 26: Weekly Average Data of recorded voltage and current density between electrodes covering seeds and plant roots

Date Range	19(A)		19(B)		Date Range	19(A)		19(B)	
	mV	*µA*	*mV*	*µA*		*mV*	*µA*	*mV*	*µA*
29.10.19 – 04.11.19	474	384	161	62	07.01.20 – 13.01.20	638	365	574	298
05.11.19 – 11.11.19	539	417	571	104	14.01.20 – 20.01.20	654	457	620	409
12.11.19 – 18.11.19	544	321	558	116	21.01.20 – 27.01.20	664	369	685	378
19.11.19 – 25.11.19	596	374	631	99	28.01.20 – 03.02.20	659	500	699	370
26.11.19 – 02.12.19	601	593	595	329	04.02.20 – 10.02.20	672	508	702	338
03.12.19 – 09.12.19	558	574	422	396	11.02.20 – 17.02.20	655	530	717	367
10.12.19 – 16.12.19	640	766	558	406	18.02.20 – 24.02.20	670	508	743	367
17.12.19 – 23.12.19	648	621	509	231	25.02.20 – 02.03.20	687	491	719	389
24.12.19 – 30.12.19	646	539	569	303	03.03.20 – 09.03.20	638	536	723	383
31.12.19 – 06.01.20	635	428	628	382					

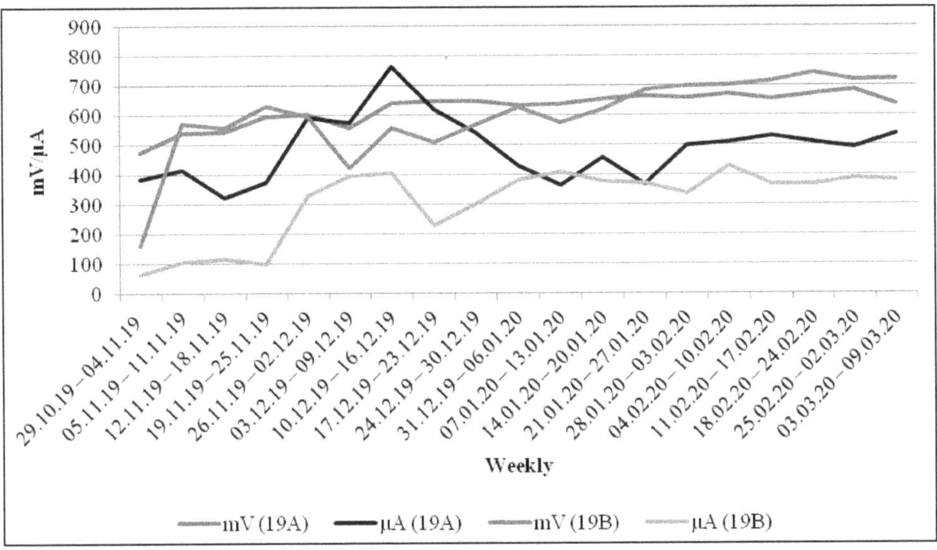

Figure 26: Pattern of Galvanic Current Flow through Seeds and Plant Roots with Two Electrode Combinations (19A – with Cu + Al and 19B - with Al + Al).

The seedlings of 19 (A), 19 (B) and control was twice transplanted from glass tank to pot and finally to the ground soil with galvanic current coverage by the same electrodes which were used during germination and their survival and growth was recorded as below.

	19 (A) (cm)	19 (B) (cm)	19 (Control) (cm)
After 20 days of germination	6.5 – 9.1	3	4
After 46 days of germination	12.2 – 18.3	1.8 – 6.4	7.5
After 54 days of germination	13.8 – 22.0	3.0 – 7.4	8.5
After 73 days of germination	34.6 – 58.0	7.0 – 17.5	24.5
After 82 days of germination	48.0 – 72.0	10.0 – 28.7	39
After 106 days of germination	94.0 – 130.0	16.0 – 68.6	129.5

After 157 days of germination the tomato fruit (green) were counted. In 19 (A) 67 fruits were counted, while there was no yield in 19 (B) and control. The highest of 6 fallen tomatoes from the plant (19A) was measured for weight, after 160 days of germination and they were between 34.0 to 51.2 gram (6 nos.). After 167 days of germination, most of the tomatoes have been ripe, and first harvest of 36 tomatoes have been made on the day from 19 (A). They were all ripe and red and their weight ranged 17.64 to 81.89 gram. Total weight of the harvest on that day was 1686 gram (1.686 Kg). The second harvest of 19 (A) was done 181 days after germination when all the 4 plants bearing fruits were in the process of drying up. During the second harvest 20 fruits (ripe) were harvested, their weight ranging from 6.89 to 6.23 gram, having a total weight of 488.46 grams. Thus from 4 plants of 19 (A) an yield of 2.194 Kg (2194 gram) was obtained with the intervention of galvanic current starting from the germination of seed and survival and growth of seedlings to mature plants bearing ripe fruits [Figures 27(A), (B), (C), (D)].

Figure 27(A)

Figure 27(B)

Figure 27(C)   **Figure 27(D)**

Applied Current Data (19 B)

Voltage applied current - 9V (9000mV)

Nature of current - PDC -1/Sec

Date	Starting Time	Stopping Time	Duration	Current Density (µA)												
				Immediate		After 1 Hour		After 2 Hour		After 3 Hour		After 4 Hour		After 13 Hour	
				mV	µA	mV	µA	mV	µA	mV	µA	mV	µA	mV	µA
29.10.19	05:30PM	06:00PM	30 min	690	320	164	30	37	0	18	0	30	0	9	0
30.10.19	04:30PM	05:30PM	60 min	627	400	471	130	332	70	170	30	75	10 *	21	0
31.10.19	05:00PM	06:30PM	90 min			499	200	444	120	355	90	277	50	8.4	0

In earlier experiments in soil and water media, soaked seeds were found to be infested with maggots of fruit fly (*Drosophila* sp.) which completely destroyed the soaked seeds resulting in failed germination in most of the cases. In order to avoid this situation, next experiments have been planned to use sterilized cotton media soaked with water and covered with bell jar. Accordingly polythene bowl of 10.5 cm upper diameter and 6.5 cm lower diameter have been used as germination containers having a depth of 5.5 cm for two experiments and one control.

The temperature of cotton media, soaked with water was 24.5°C. Bent aluminium and copper plate fitting to the shape of bowl and separated at a distance 8.5 cm at the outer rim and 4.5 cm at the bottom, placed vertically served as electrodes for 20 (A); which in 20(B) two bent aluminium plates of same size have been fixed in the same manner as that of 20(A).

8 tomato seeds (*Solanum lycopersicum*), 10 green broad bean (*Vicia faba*) and 8 ash gourd (*Benincara hispida*) were placed on the surface of soaked cotton in each of the experimental bowl and in control.

The galvanic current characteristic at the start was 638 mV and 360 µA in 20(A) and 10.7 mV and 0 µA in 20(B). For next two days 594 mV and 524 mV along with 490 µA and 780 µA prevailed in 20(A), while in 20(B), the values were 115 mV and 130 mV along with 0 µA in both the days.

On the third day, that is, after 72 hours, 7 out of 8 tomato seeds(87.5 per cent), 7 out of 10 green broad bean (70 per cent) and 2 out of 8 ash gourd (25 per cent) seeds germinated in 20(A). In 20(B), during the period, 87.5 per cent tomato, 30 per cent green broad bean only germinated. In control 75 per cent tomato, 60 per cent bean and 25 per cent ash gourd germinated without any influence of galvanic current. Within 96 hours another 10 per cent bean germinated in 20(A); another 20 per cent bean and 25 per cent ash gourd germinated in 20(B). In control however, one germinating tomato seed died, another 20 per cent bean and 25 per cent ash gourd germinated in control.

Most of the soaked bean seeds were found infested with maggots of fruit fly which entered in the bell jar through a small gap at the bottom of the bell jar, which has been plugged with cotton. Surprisingly there was no infestation in 20(A) through the bowl with Al and Cu electrode was in the same bell jar with 20(B). This was possibly due to the presence of copper electrode in 20(A), which might have kept the fruit fly away to lay eggs in 20(A) bowl and inhibited maggot to grow.

After 3 days of germination the following seedlings have been survived and grown in experimental bowls and control.

In 20(A) – Tomato – (100 per cent, 0.8 – 6 cm); Bean (80 per cent, 1.3 – 13 cm); Ash gourd (12.5 per cent, 1 cm)

In 20(B) – Tomato – (100 per cent, 2.7 – 5.5 cm); Bean (50 per cent, 3 – 10.5 cm); Ash gourd (12.5 per cent, 0.2 cm)

Control – Tomato – (75 per cent, 4.1 – 4.8 cm), Bean (80 per cent, 2.8 – 7.5 cm), Ash gourd (37.5 per cent, 0.3 – 0.8 cm)

All the tomato and bean seedlings were transferred in pots with earthen media after taking the measurements of growth. They were not provided with electrodes on that day. Only ash gourd remained in bowl in cotton media with electrodes in 20(A) and 20(B) and without any electrode in control.

Next day after shifting tomato seedlings in earthen pots, electrodes were introduced in the pots of 20(A) and 20(B) as was in the experimental bowls, for germination of seeds Cu + Al in 20(A) and Al + Al in 20(B).

Table 28: Weekly Average Data of Recorded Voltage and Current Density between Electrodes Covering Seeds and Plant Roots (Tomat)

Date Range	20(A)		20(B)		Date Range	20(A)		20(B)	
	mV	μA	mV	μA		mV	μA	mV	μA
10.11.19 – 16.11.19	553	445	73	45.94	09.02.20 – 15.02.20	624	270	53.4	1.67
17.11.19 – 23.11.19	571	263	22	0	16.02.20 – 22.02.20	646	233	49	5
24.11.19 – 30.11.19	624	470	17.3	1.43	23.02.20 – 29.02.20	630	337	6.2	0
01.12.19 – 07.12.19	628	389	22	1.43	01.03.20 – 07.03.20	621	360	7.7	0
08.12.19 – 14.12.19	572	259	42	7.14	08.03.20 – 14.03.20	671	320	59.4	0
15.12.19 – 21.12.19	531	199	90	15.7	15.03.20 – 21.03.20	668	329	37.1	0
22.12.19 – 28.12.19	539	194	138	5.7	22.03.20 – 28.03.20	678	327	28	0
29.12.19 – 04.01.20	562	203	46.5	0	29.03.20 – 04.04.20	702	252	52	2
05.01.20 – 11.01.20	540	124	20	4	05.04.20 – 11.04.20	674	297	107	6.67
12.01.20 – 18.01.20	580	177	34	1.6	12.04.20 – 18.04.20	760	254		
19.01.20 – 25.01.20	661	240	40	3.3	19.04.20 – 25.04.20	685	198		
26.01.20 – 02.02.20	599	313	13.3	1.43	26.04.20 – 02.05.20	638	176		
03.02.20 – 08.02.20	615	269	12.2	0	03.05.20 – 06.05.20	612	180		

After 9 days of shifting seedlings of tomato into the pot, all the tomato seedlings and seedlings of bean and ash gourd were transplanted into ground soil with copper and aluminium electrodes around the seedlings in the ground soil for 20(A) and aluminium electrodes around the seedlings in the ground soil for 20(B) and no electrodes around the seedlings of the control. Separate sets of electrodes have been put in the ground soil for tomato, bean and ash gourd. During 182 days in the ground soil, mV and μA around tomato seedlings were fluctuating between 256 – 771 mV and 22 – 660 μA in 20(A). Likewise in 20(B) 1 – 20 mV and 0 – 240 μA were prevailing during those days. Calculating a weekly average, for tomato in 20(A), the range was 531 – 760 mV and 124 – 470 μA, whereas, for 20(B), the variation was 6.2 – 138 mV and 0 – 45.94 μA.

In case of bean seedlings on ground soil, the electrodes around their roots delivered 237 – 715 mV and 22 – 780 μA in 20(A) and 2 – 568 mV and 0 – 660 μA in 20(B). Their weekly variations were calculated as 452 – 662 mV and 53 – 445 μA in 20(A) and 7.2 – 204 mV and 0 – 45.95 μA in 20(B).

Galvanic potential of 393 – 680 mV and intensity of 22 – 980 μA prevailed around the roots of ash gourd seedlings in ground soil of 20(A), while in 20(B) they were between 5 – 170 mV and 0 – 20 μA. The weekly average was 553 – 661 mV and 349 – 703 μA around 20(A) and 11 – 73 mV and 0 – 45.95 μA around 20(B). (Table 28, Figure 28).

The growth of tomato seedlings between 16.01.19 to 26.01.20 (375 days) were from 0.8 – 6 cm to 22.86 – 63.5 cm in 20(A); from 2.7 – 5.5 cm to 45.72 – 91.44 cm in 20 (B) and from 4.1 – 4.8 cm to 22.86 – 49.53 cm in control. The growth of tomato seedlings was highest in 20 (B) followed by 20 (A) and the lowest was in control. The survival of tomato seedlings were highest in both 20 (A) (75 per cent), and in 20(B) where it was 75 per cent and lowest in control (37.5 per cent).

Bean seedlings between 16.01.19 to 5.12.19 (324 days) grew from 1.3 – 1.0 cm to 9.2 – 92 cm in 20 (A); from 3.0 – 10.5 cm to 27.3 – 75 cm in 20 (B) and 2.8 – 7.5 cm to 30 – 78 cm in control indicating highest growth in 20 (A) followed by control and lowest in 20 (B). As regards the survival of plants, only 40 per cent survival (4 plants) in 20 (B) was noticed as against 80 per cent survival of plants in 20 (A) and control.

Ash gourd seedlings were reared from 16.01.19 to 02.01.20 (351 days). Their growth varied from 0.8 – 1 cm to 4.3 – 11.8 cm in 20(A); from 0.2 – 0.8 cm to 5.6 – 8.7 cm in 20 (B) and from 0.3 – 08 cm to 4.7 – 9.5 cm in control. Highest growth was recorded in 20 (A) followed by control and lowest in 20 (B). Seventy five percent survivals of ash gourd plants were noticed in both 20 (A) (75 per cent) and control while lowest 37.5 per cent was observed in 20 (B).

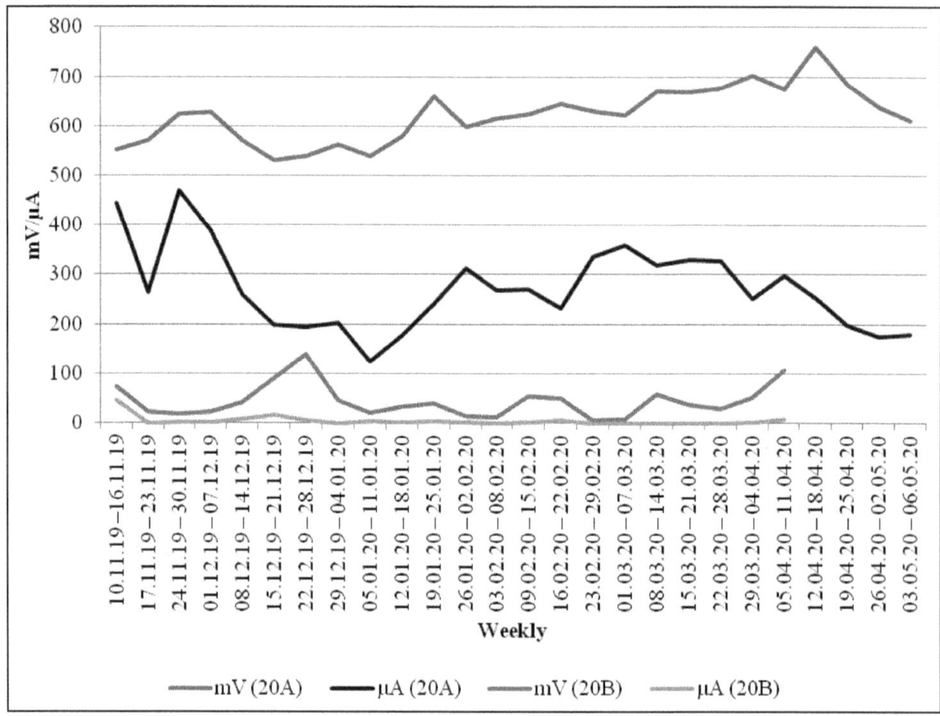

Figure 28: Pattern of Galvanic Current Flow through Seeds and Plant Roots with Two Electrode Combinations (20A – with Cu + Al and 20B - with Al + Al).

Thus it is evident that galvanic current does have a beneficial effect (with respect of higher percentage of germination, increased growth and higher percentage of survival) at least in case of tomato, beans and ash gourd as observed in the present study.

148 days after germination ripe tomatoes which turned into bright yellow were harvested from 20 (A) and 20 (control) on 11/4/20. A total of 26 tomatoes weighing 751.03 gram (6.02 – 58.76 g each) were harvested from 20 (A), while 16 tomatoes weighing 222.47 gram (1.46 – 30.95 gram each) was harvested from control.

☆ **20 (A) - Experimental** – Parabolic copper and aluminium of 12 X 5 cm size made of 18 gauge plate, inserted in the ground soil around the plants. The distance between the electrodes was 11 – 13 cm. The distance between plants and electrodes was 0.5 – 2.5 cm and the distance between plants was 1 – 2 cm. plants have attained a maximum length of 193 – 195 cm having 9 – 13 branches in each plant. They bore 26 ripe tomatoes weighing between 6.02 – 58.76 g each with a total weight of 751.03 g. Weight of the biomass (foliage) ranged from 280 – 325 g.

☆ **20 (B) – Experimental** – Parabolic two aluminium plates of 10 X 5 cm size made of 18 gauge plate, inserted in the ground soil around the plants. The distance between the electrodes was 10 – 12 cm. The distance between plants and electrodes was 0.5 – 3 cm and the distance between plants was 1 – 2 cm. Maximum length of plants were 180 – 190 cm with 3 – 7 branches. No fruiting. Weight of the biomass (foliage) ranged from 100 – 350 g.

☆ **20 (C) – Control** – Plants were put in ground soil at a distance of 3 – 4 cm between them. Attained a maximum length of 170 – 240 cm with 6 – 7 branches in each. 16 ripe fruits have been harvested, weighing between 1.46 – 30.95 gm and total weight of 222.47 g. Weight of biomass (foliage) was 250 – 350 g each.

Effect of Galvanic Current to Seedlings

To study the effect of galvanic current on egg plant seedling (*Solanum melongena*) in earthen media in a pot provided with copper and aluminium electrodes to produce galvanic current one 15 cm long seedling with 3 leaves was planted in the pot along with a control in a separate pot, where a 18 cm seedling with 4 leaves was introduced. The galvanic current for the first 19 days around the seedling were from 221 – 679 mV and 270 – 1230 µA. But the seedling died and the replacement was made with that of control one and a new seedling (16 cm with 3 leaves) was put as control. After replacement of the seedling in experimental pot, the galvanic current characteristics were found to be 306 – 703 mV and 250 – 1110 µA and the weekly average was 419 – 683 mV and 396 – 986 µA [Table 29, Figures 29(A), (b)].

Table29 (A): Weekly Average Data of Recorded Voltage and Current Density between Electrodes Covering Plant Roots

Date Range	mV	µA	Date Range	mV	µA
09.03.20 – 15.03.20	419	543	13.04.20 – 19.04.20	683	986
16.03.20 – 22.03.20	330	433	20.04.20 – 26.04.20	651	861
23.03.20 – 29.03.20	524	566	27.04.20 – 03.05.20	630	910
30.03.20 – 05.04.20	543	396	04.05.20 – 06.05.20	549	620
06.04.20 – 12.04.20	607	651			

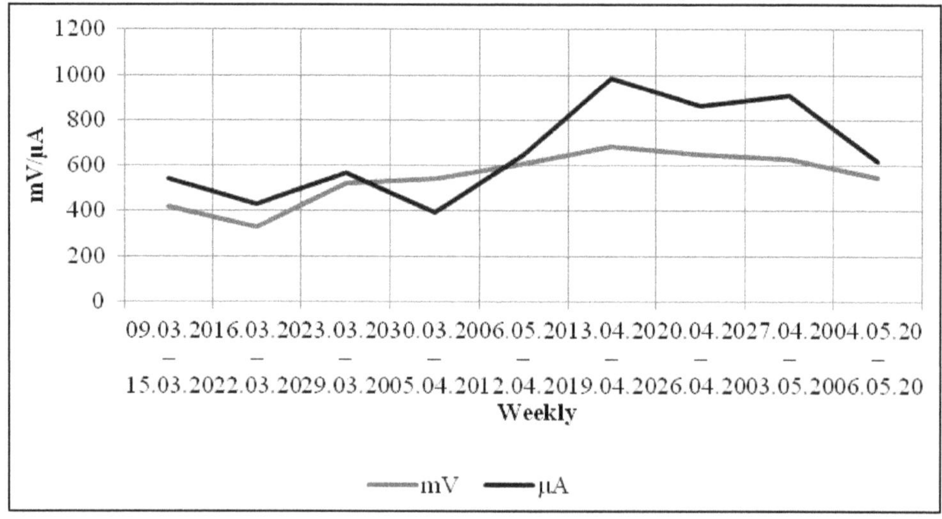

Figure 29(A): Pattern of Current Flow through Plant Roots with Cu + Al Electrodes

Table 29(B): Weekly Average Data of Recorded Voltage and Current Density between Electrodes Covering Plant Roots

Date Range	mV	μA	Date Range	mV	μA
22.06.20 – 28.06.20	606	634	13.07.20 – 19.07.20	483	580
29.06.20 – 05.07.20	624	663	20.07.20 – 26.07.20	622	733
06.07.20 – 12.07.20	596	731	27.07.20 – 02.08.20	567	547

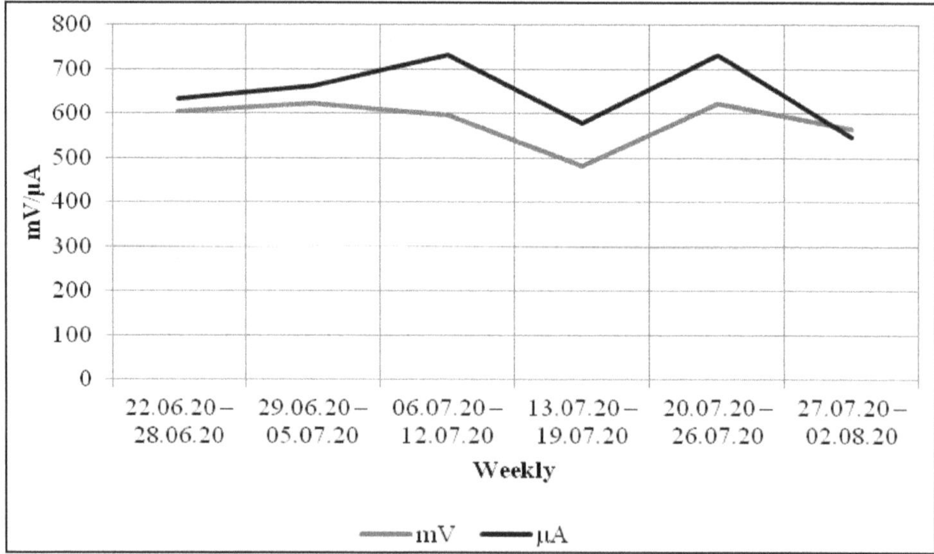

Figure 29(B): Pattern of Current Flow through Plant Roots with Cu + Al electrodes (After replacement of control).

Within 11 days the seedling of experimental pot attained a length of 26.5 cm, an increment of 11.5 cm with 7 leaves. The control however did not survive. Within a month, the plant developed 3 branches of 23 cm, 28 cm and 42 cm long with a dozen of leaves. Within 38 days, flowering started 2 – 3 at a time, but fell off within a week. During second flowering, which appeared after 47 days, out of 3 flowers, one flower bore a fruit, one at a time. The fruit grew to an edible size of 148 gm in another 12 days and was harvested. The second flowering continued and within a week of first harvest, another flower developed into fruit and this time also, one at a time. After 20 days, the second fruit had attained edible size of 143 gm and was harvested.

After the second fruiting the plant root and stem was infested with ants, the leaves became yellow and fell off resulting in death of the plants.

Though the findings of the experiment was not very conclusive, because of the absence of control one, which died earlier, yet, the beneficial effect of galvanic current around the plant in bearing flowers and fruits in a pot cannot be ignored.

In another experiments, one 4.5 cm bottle gourd seedling (*Lagenaria siceraria*) was subjected to galvanic current exposure in the ground soil with the help of aluminium and copper electrodes inserted in the ground at a distance of 4 cm on each side of the seedling and their growth was measured after 37 days.

During the period seedling was exposed to induce galvanic current (both potential difference and current intensities) round the clock for 37 days. The potential difference to which the seedling was exposed ranged from 508 – 712 mV and the current intensity of 160 – 570 µA (recorded from daily observations). The weekly average of recorded data of voltage and current intensity was (Table 30, Figure 30).

Table 30: Weekly Average Data of Recorded Voltage and Current Density

Week No.	Date Range	mV	µA	Week No.	Date Range	mV	µA
1st	14.02.20 – 20.02.20	613	319	7th	27.03.20 – 02.04.20	628	317
2nd	21.02.20 – 27.02.20	586	323	8th	03.04.20 – 09.04.20	672	293
3rd	28.02.20 – 05.03.20	603	260	9th	10.04.20 – 16.04.20	567	224
4th	06.03.20 – 12.03.20	594	217	10th	17.04.20 – 23.04.20	623	229
5th	13.03.20 – 19.03.20	634	300	11th	24.04.20 – 30.04.20	603	206
6th	20.03.20 – 26.03.20	629	296	12th	01.05.20 – 06.05.20	623	300

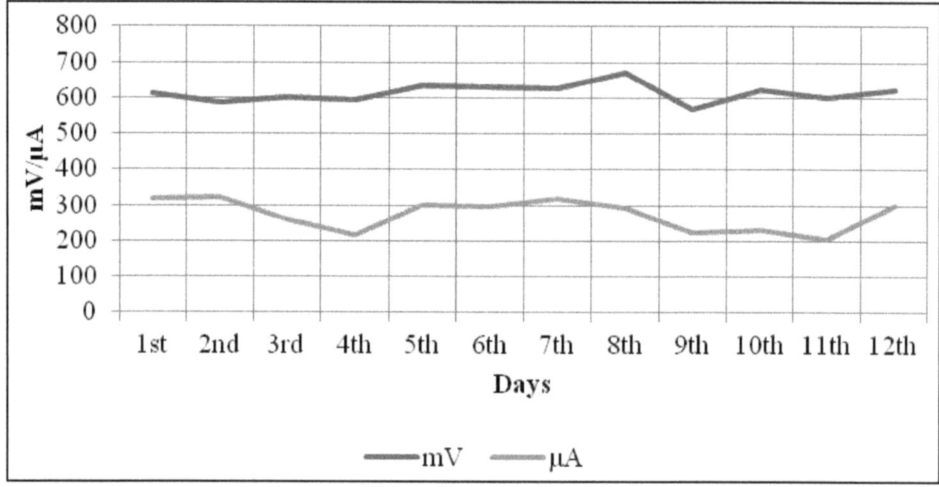

Figure 30: Pattern of Galvanic Current Flow through Roots of Seedlings.

The growth of the seedling was measured for the length, and found the plant had grown to 14.4 cm on 20.03.20 (an increment of 9.9 cm) and 27.3 cm on 06.05.20 (an increment of 12.9 cm in 47 days). Control could not be run due to non – availability of seedling.

Another trial with bottle gourd along with the control was taken up on 02.03.20 in ground soil with copper and aluminium electrodes for the experimental one. The initial length of the seedlings were 3.5 cm both for experiment and control. The test lasted for 66 days. During the period the galvanic current exposure to the experimental seedling was 539 – 665 mV and 130 – 570 µA. The weekly average of voltage and current intensity; were; 581 mV and 272 µA for the first week; 616 mV and 284 µA for the second week; 626 mV and 341 µA for the third week; 639 mV and 339 µA for the fourth week; 650 mV and 284 µA for the fifth week; 590 mV and 217 µA for the sixth week; 599 mV and 254 µA for the seventh week; 608 mV and 220 µA for the eighth week; 584 mV and 186 µA for the ninth week and 601 mV and 180 µA for the tenth week (Table 31, Figure 31).

Table 31: Weekly Average Data of Recorded Voltage and Current Density

Week No.	Date Range	mV	µA	Week No.	Date Range	mV	µA
1st	02.03.20 – 08.03.20	581	272	6th	06.04.20 – 12.04.20	590	217
2nd	09.03.20 – 15.03.20	616	284	7th	13.04.20 – 19.04.20	599	254
3rd	16.03.20 – 22.03.20	626	341	8th	20.04.20 – 26.04.20	608	220

Week No.	Date Range	mV	μA	Week No.	Date Range	mV	μA
4th	23.03.20 – 29.03.20	639	339	9th	27.04.20 – 03.05.20	584	186
5th	30.03.20 – 05.04.20	650	284	10th	04.04.20 – 06.05.20	601	180

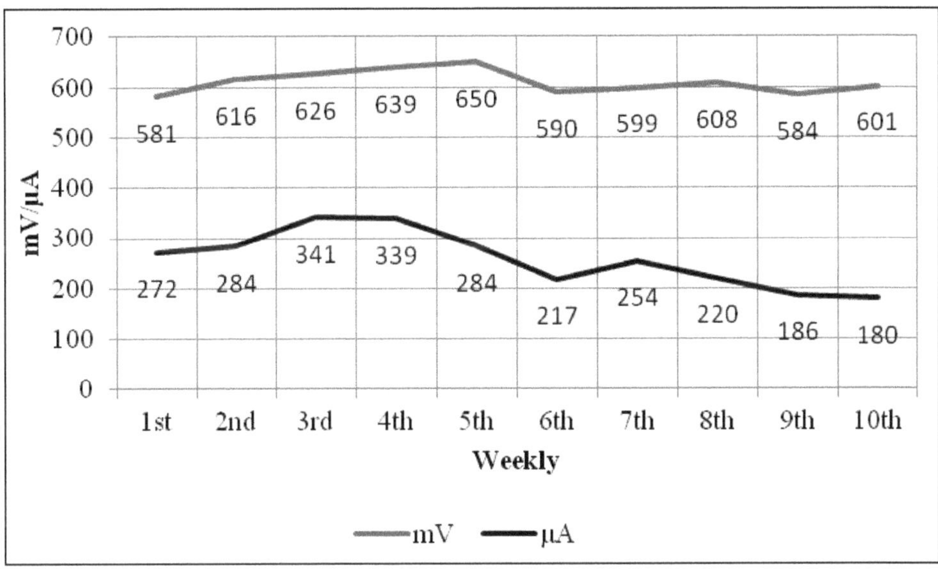

Figure 31: Pattern of Galvanic Current Flow through Roots of Seedlings.

The experimental plant had grown to a length of 14.4 cm in 18 days, an increment of 9.9 cm in 18 days. The control seedling did not survive.

Effect of Galvanic Current on Rhizomes (*Canna indica*)

Three sets of Canna rhizomes were placed under the exposure of galvanic current in soil media, first in the pot and subsequently in the ground soil, along with control to study the sprouting and growth of the flowering plant. All the three experimental sets were provided with the same type of electrodes (Copper and aluminium).

The range of recorded voltage and current density generated in the three sets were;

	At 24 Hours Interval		Weekly Average		Electrodes		
	mV	*µA*	*mV*	*µA*	*Anode*	*Cathode*	
First set	178 – 710	80 – 630	509 – 631	210 – 440	Copper	Aluminium	[Tables 33(A), (B); Figures 33(A), (B)]
Second set	322 – 955	150 – 1000	613 – 910	639 – 1413	Copper	Aluminium	[Tables 33(A), (B); Figures 33(A), (B)]
Third set	198 – 665	30 – 1170	463 – 586	326 – 899	Copper	Aluminium	[Tables 33(A), (B); Figures 33(A), (B)]

Table 32(A): Weekly Average Data of Recorded Voltage and Current Density between Electrodes Covering Plant Rhizome

Week No.	Date Range	mV	µA	Week No.	Date Range	mV	µA
1st	09.03.20 – 15.03.20	577	396	6th	13.04.20 – 19.04.20	566	364
2nd	16.03.20 – 22.03.20	562	321	7th	20.04.20 – 26.04.20	559	403
3rd	23.03.20 – 29.03.20	574	327	8th	27.04.20 – 03.05.20	547	358
4th	30.03.20 – 05.04.20	586	380	9th	04.05.20 – 06.05.20	568	440
5th	06.04.20 – 12.04.20	545	210				

Table 32(B): Weekly Average Data of Recorded Voltage and Current Density between Electrodes Covering Plant Rhizome

Week No.	Date range	mV	µA	Week No.	Date range	mV	µA
1st	22.06.20 – 28.06.20	530	413	4th	13.07.20 – 19.07.20	631	371
2nd	29.06.20 – 05.07.20	578	391	5th	20.07.20 – 26.07.20	512	318
3rd	06.07.20 – 12.07.20	509	424				

Table 33(A): Weekly Average Data of Recorded Voltage and Current Density between Electrodes Covering Plant Rhizome

Week No.	Date range	mV	µA	Week No.	Date range	mV	µA
1st	10.03.20 – 16.03.20	910	1413	6th	14.04.20 – 20.04.20	703	711
2nd	17.03.20 – 23.03.20	898	1031	7th	21.04.20 – 27.04.20	778	801
3rd	24.03.20 – 30.03.20	849	863	8th	28.04.20 – 04.05.20	846	860
4th	31.03.20 – 06.04.20	831	828	9th	05.05.20 – 06.05.20	838	990
5th	07.04.20 – 13.04.20	680	639				

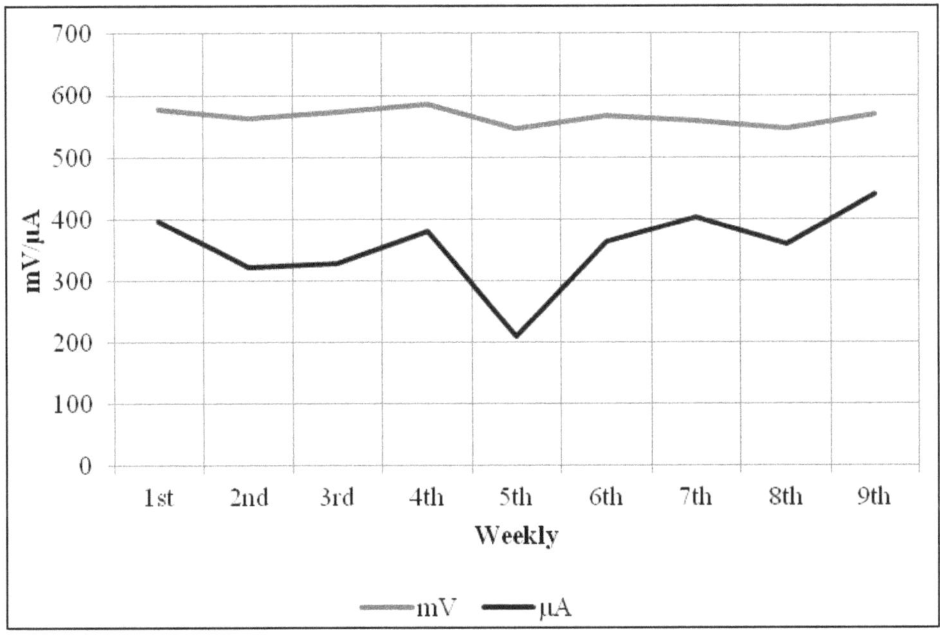

Figure 32(A): Pattern of Galvanic Current Flow through Rhizome with Cu + Al Electrodes.

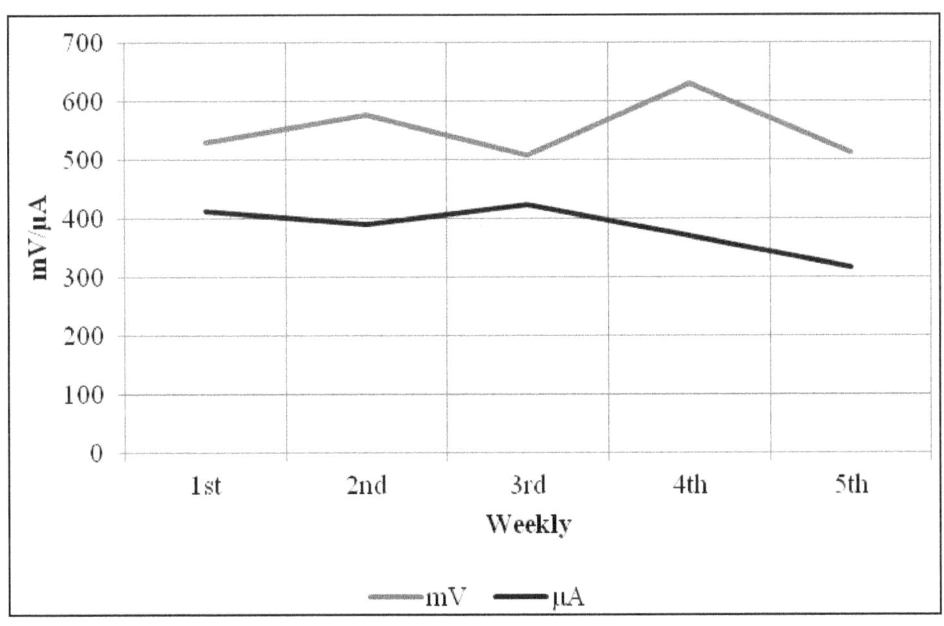

Figure 32(B): Pattern of Galvanic Current Flow through Rhizome with Cu + Al Electrodes.

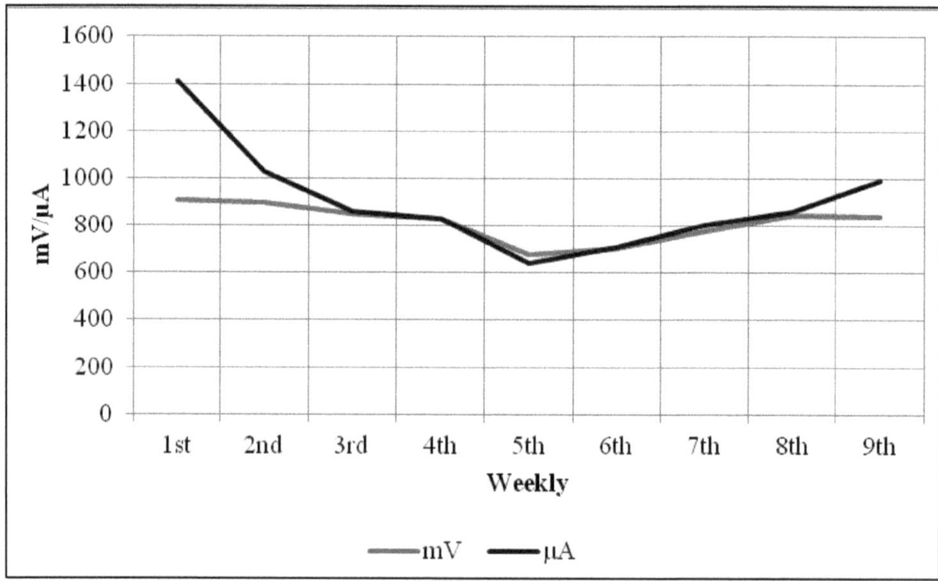

Figure 33(A): Pattern of Galvanic Current Flow through Rhizome with Cu + Al Electrodes.

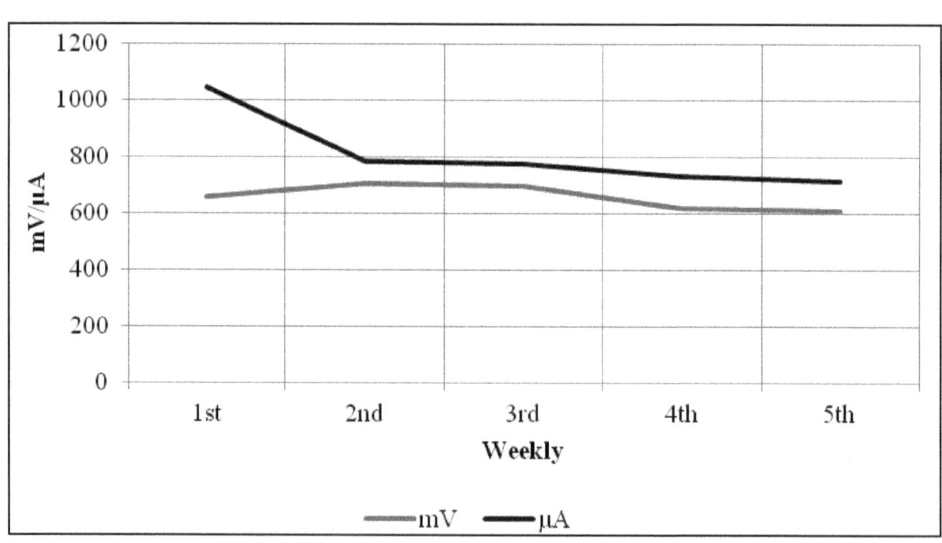

Figure 33(B): Pattern of Galvanic Current Flow through Rhizome with Cu + Al Electrodes.

Table 33(B): Weekly Average Data of Recorded Voltage and Current Density between Electrodes Covering Plant Rhizome

Week No.	Date range	mV	µA	Week No.	Date range	mV	µA
1st	22.06.20 – 28.06.20	660	1047	4th	13.07.20 – 19.07.20	621	736
2nd	29.06.20 – 05.07.20	707	784	5th	20.07.20 – 25.07.20	613	715
3rd	06.07.20 – 12.07.20	698	777				

Table 34(A): Weekly Average Data of Recorded Voltage and Current Density between Electrodes Covering Plant Rhizome

Week No.	Date range	mV	µA	Week No.	Date range	mV	µA
1st	17.03.20 – 23.03.20	516	326	5th	14.04.20 – 20.04.20	570	769
2nd	24.03.20 – 30.03.20	564	560	6th	21.04.20 – 27.04.20	572	899
3rd	31.03.20 – 06.04.20	558	698	7th	28.04.20 – 04.05.20	580	865
4th	07.04.20 – 13.04.20	567	736	8th	05.05.20 – 06.05.20	582	800

Table 34(B): Weekly Average Data of Recorded Voltage and Current Density between Electrodes Covering Plant Rhizome

Week No.	Date range	mV	µA	Week No.	Date range	mV	µA
1st	22.06.20 – 28.06.20	463	649	4th	13.07.20 – 19.07.20	534	553
2nd	29.06.20 – 05.07.20	479	717	5th	20.07.20 – 25.07.20	566	723
3rd	06.07.20 – 12.07.20	586	814				

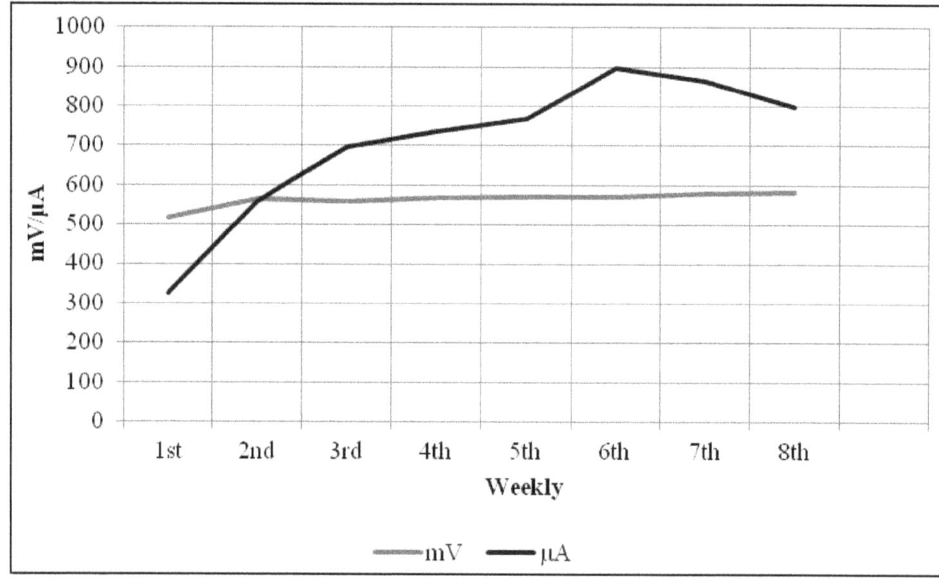

Figure 34(A): Pattern of Galvanic Current Flow through Rhizome with Cu + Al Electrodes.

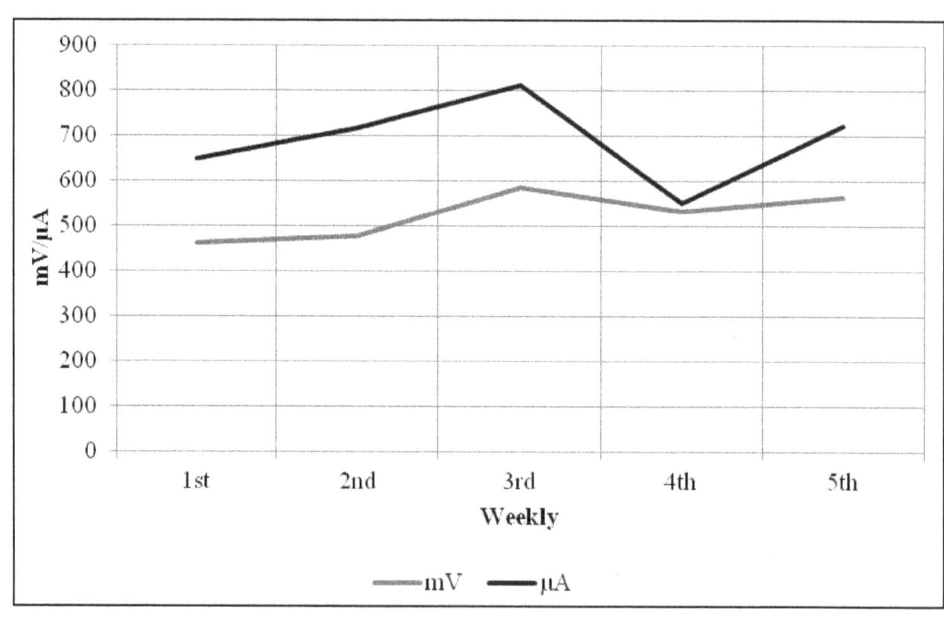

Figure 34(B): Pattern of Galvanic Current Flow through Rhizome with Cu + Al Electrodes.

The plant in first set had grown from 7 cm to 19 cm in 11 days while, the length of control plant was 10 cm.

In the second set the plant had grown from 8 cm to 16.5 cm in 10 days and the control attained a length of 11.1 cm from 7.5 cm during the period.

The length of the plant in third set was measured as 12.5 cm on the third day, when the length of control plant was 13 cm.

The first flowering of experimental plant in the first set was 32 days earlier than the plant in control. The flowering (first) of second set was 22 days in advance of the plant in control. The second flowering of the second set was after 76 days of first flowering, while there was no sign of flowering in the control. In the third set flower appeared 34 days earlier than the control plant.

The growth in length of plants treated with galvanic current was invariably more than the plants which grew without treatment. The flowering in all the three experimental sets, exposed to galvanic current appeared 22 to 34 days earlier than their respective control counter parts. Even the second flowering appeared in the second, first and third sets when there was no sign of appearance of buds in their respective control counter parts.

The study, therefore, reveals some beneficial effect with respect to growth and appearance of flowers in galvanically treated plants in comparison with their control counter parts.

12

Effect of Galvanic Current on the Root Formation of Buds

Cycas (*Cycas revoluta*) buds, collected from a cycas plant, without any root formation was put in plastic pots in soil media, with aluminium and copper electrodes in one pot and without any electrode in the control one. The pot with electrodes (Exptl one) was provided two buds, 30 mm and 18 mm in diameters at the base, while in control pot, one bud with 34 mm diameter was planted in the soil.

After 6 days, the larger bud in experimental pot was found to form roots and a growing leaf; while the smaller bud in the same pot and the bud in the control pot had no sign of root and leaf formation. The leaf of the cycas bud started growing rapidly in the beginning (about more than 10 mm daily) and became 141 mm long within a month. The smaller bud in the experimental pot had an outshoot of leaf after 30 days from the bigger one. In control pot, with a bigger bud has exhibited the growth of an out shoot of leaf after 15 days from the experimental pot.

The galvanic current flowing at the bases of cycas buds fluctuated from 460 – 810 mV and 190 – 550 µA (Table 35, Figure 35).

Table 35: Galvanic Current Characteristics Beneath the Cycas Buds in the Experimental Pot

Date	Exp - I		Date	Exp - I		Date	Exp - I	
	mV	*µA*		*mV*	*µA*		*mV*	*µA*
22.08.19	810	370	28.08.19	673	320	03.09.19	537	300
23.08.19	685	440	29.08.19	725	450	04.09.19	614	300

Date	Exp - I		Date	Exp - I		Date	Exp - I	
	mV	*µA*		*mV*	*µA*		*mV*	*µA*
24.08.19	617	340	30.08.19	753	550	05.09.19	627	330
25.08.19	624	300	31.08.19	732	450	06.09.19	513	220
26.08.19	477	190	01.09.19	596	375	07.09.19	709	320
27.08.19	621	190	02.09.19	460	300	08.09.19	593	230

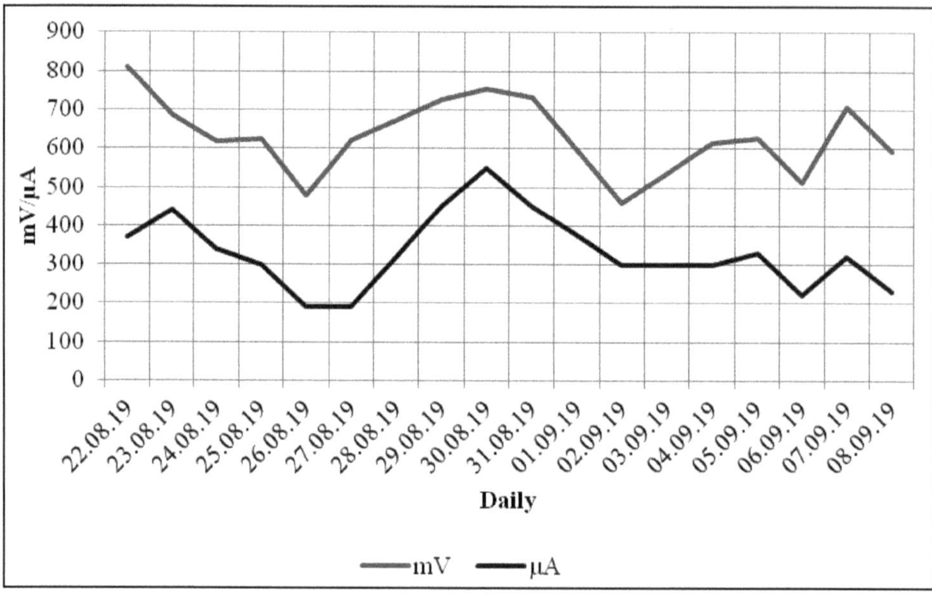

Figure 35: Pattern of Galvanic Current Flow through the Base of Cycas Buds of Experimental Pot.

Galvanic current flowing through the soil at the base of the cycas buds have been found to have a beneficial effect on the early formation of roots and shoots (leaf).

Influence of Galvanic Current on the Direction of Root's Growth

To study formation and propagation of roots in galvanic current field, two ladies finger (*Abelmoschus esculentus*) seedlings were planted at center point (middle of the glass tank containing ground soil) in order to observe the direction and growth of roots. One copper and one GI plates were inserted into the ground soil at the longest extremities of the tank to produce galvanic current.

The galvanic current characteristics of the glass tank ranged from 557 – 940 mV and 306 – 980 µA between 11.03.20 to 06.05.20 and 363 – 682 mV and 8.6 – 1045 µA between 22.06.20 to 03.09.20 [Tables 36(A), (B); Figures 36(A), (B)].

Table 36(A): Weekly average data of recorded voltage and current density influencing propagation of roots

Week No.	Date Range	mV	μA	Week No.	Date Range	mV	μA
1st	11.03.20 – 17.03.20	940	873	5th	08.04.20 – 14.04.20	591	753
2nd	18.03.20 – 24.03.20	935	980	6th	15.04.20 – 21.04.20	667	831
3rd	25.03.20 – 31.03.20	841	812	7th	22.04.20 – 28.04.20	557	377
4th	01.04.20 – 07.04.20	678	818	8th	29.04.20 – 06.05.20	569	306

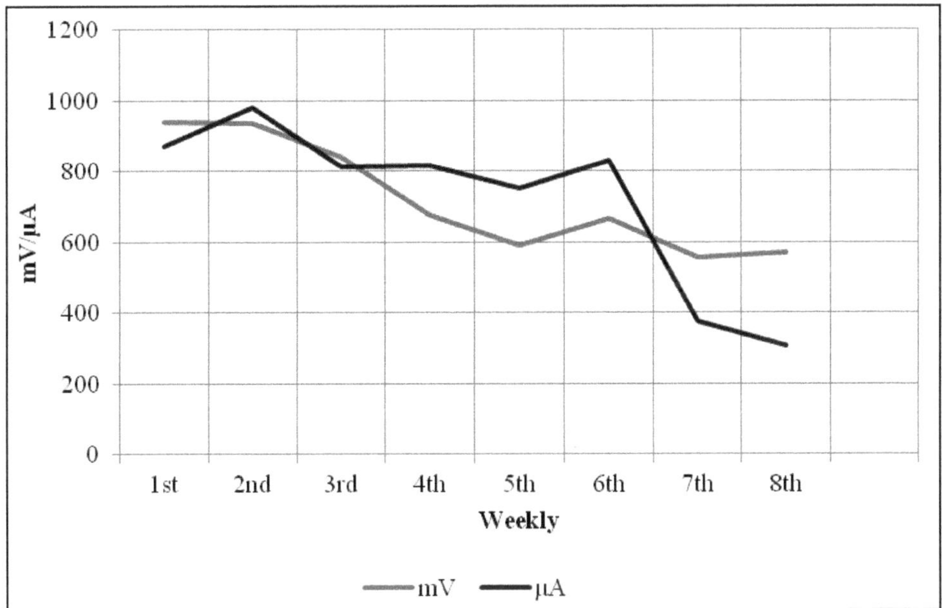

Figure 36(A): Pattern of Galvanic Current Flow through Roots of Seedlings.

During the period, both the plants grew to 8.6 and 8.9 cm (Initial length of 4.5 and 6 cm) in length in 19 days, and 29 – 36 cm and 43 – 52 cm in length on 127 days and 197 days respectively. The plant near negative electrode was found 14 cm taller than the plant near positive electrode. The roots of the plant near positive electrode were found to be 5.4 – 9.8 cm long while that of plant near negative electrode, were 4.8 – 6.2 cm long. 73 per cent of plant roots situated near the positive electrode were turning and growing towards the copper plate (positive electrode), while 27 per cent of the plant root were pointing towards the negative electrode (galvanized

Table 36(B): Weekly Average Data of Recorded Voltage and Current Density Influencing Propagation of Roots

Week No.	Date Range	mV	µA	Week No.	Date Range	mV	µA
1st	22.06.20 – 28.06.20	363	8.6	8th	03.08.20 – 09.08.20	638	537
2nd	29.06.20 – 05.07.20	389	118	9th	10.08.20 – 16.08.20	682	644
3rd	06.07.20 – 12.07.20	614	437	10th	17.08.20 – 23.08.20	458	700
4th	13.07.20 – 19.07.20	600	349	11th	24.08.20 – 30.08.20	548	887
5th	20.07.20 – 26.07.20	646	549	12th	31.08.20 – 03.09.20	618	1045
6th	27.07.20 – 02.08.20	609	562				

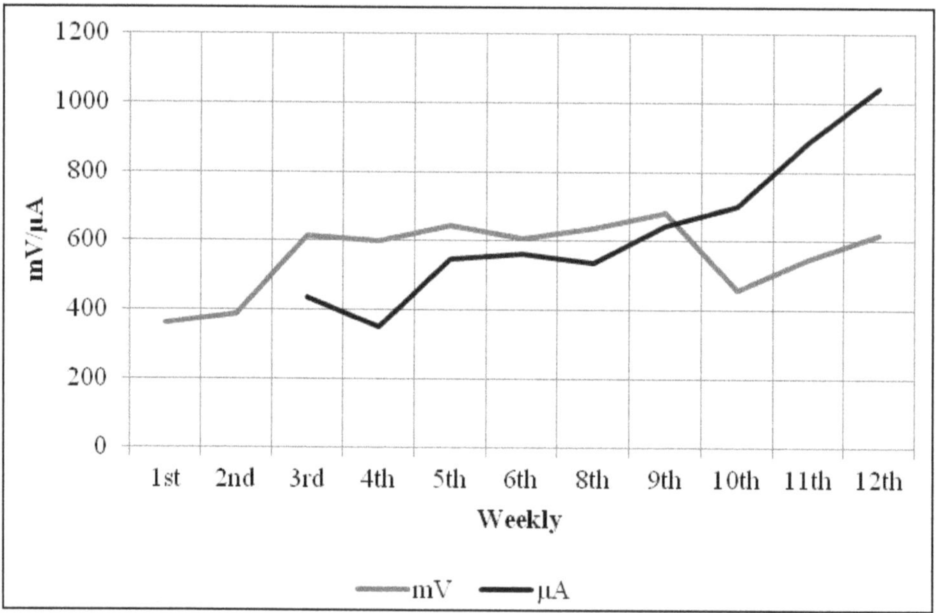

Figure 36(B): Pattern of Galvanic Current Flow through Roots of Seedlings.

iron plate). The roots of plant near negative electrode were found less in number (11) while the plant at positive electrode was having 18 roots of which 45 per cent turned and growing towards the +Ve electrode. The rest 55 per cent of the roots were growing towards –Ve electrode (Figure 37).

Figure 37: Growth of Ladies Finger Seedlings (Left); the Roots of the Seedling were Growing towards the Anode (Right).

C
h
a
p
t
e
r

13

Comparative Effectiveness of Electrode Combinations

In earlier chapters mention have been made about how the generation of galvanic current is influenced with nature, size and distance between two electrodes. It has also been mentioned that the intensity of galvanic current is more; when electrodes of two dissimilar metals are used than electrodes of the same metal. The observations were based on the measuring the current intensity (mV and μA) by precision measuring instruments.

In order to have a comparative study in the field (both in pot soil and ground soil), 6 papaya (*Carica papaya*) seedlings (15.0 – 16.3 cm) were put under galvanic current exposure (one in pot soil and 5 in ground soil) with different electrode combinations as detailed below.

☆ **33 (A)** – In pot soil with Al and Cu electrodes. The distance between two electrodes was 17 cm and from seedling 7 cm from Cu and 10 cm from Al electrodes.

☆ **33 (B)** – In ground soil with Zn and Al electrodes. The distance between two electrodes was 23 cm and from seedling 9 cm from Zn and 14 cm from Al electrodes.

☆ **33 (C)** – In ground soil with Cu and Al electrode. The distance between two electrode was 23 cm and from seedling 13 cm from Cu and 10 cm from Al electrodes.

☆ **33 (D)** – In ground soil with Cu and GI electrodes. The distance between two electrodes was 12.5 cm and from seedling 5 cm from Cu and 7.5 cm from GI electrodes.

☆ **33 (E)** – In ground soil with Al and Al electrodes. The distance between two electrodes was 16 cm and from seedling 8 cm from each of the Al electrodes.

☆ **33 (F)** – In ground soil with Cu and Cu electrodes. The distance between two electrodes was 13 cm and from seedlings 6.5 cm from each of the Cu electrode.

Besides two control, one in pot soil and the other in ground soil were run without any electrodes in identical environment.

The experiments started on 23.6.20 and continue till 18.09.20. The weekly average data of recorded voltage and current density with effect from 23.6.20 to 18.9.20 indicated a fluctuation of 488 – 999 mV and 207 – 1160 µA in 33 (A); 445 – 618 mV and 211 – 603 µA in 33 (B); 373 – 662 mV and 116 – 370 µA in 33 (C); 644 – 943 mV and 151 – 2182 µA in 33 (D); 8.5 – 115 mV and 0 – 57.1 µA in 33 (E) and 27 – 314 mV and 2.5 – 45 µA in 33(F) (Table 37, Figure 38).

Table 37: Weekly Average Data of Recorded Voltage and Current Density Flowing through Roots of Seedlings

Week No.	Date	33(A)		33(B)		33(C)		33(D)		33(E)		33(F)	
		mV	µA	mV	µA	mV	µA	mV	µA	mV	µA	mV	µA
1st	23.06.20 – 29.06.20	804	797	445	211	579	310	829	2025	44.2	57.1	231	45
2nd	30.06.20 – 06.07.20	529	291	552	327	587	337	804	1204	72	13	71	14
3rd	07.07.20 – 13.07.20	523	336	492	319	565	347	851	1414	22	4.3	307	4.3
4th	14.07.20 – 20.07.20	488	207	589	276	543	363	795	151	115	10	295	9
5th	21.07.20 – 27.07.20	624	525	483	320	594	370	828	1847	47	10	97	10
6th	28.07.20 – 03.08.20	627	500	520	367	409	163	834	1457	63.1	21	185	27
7th	04.08.20 – 10.08.20	860	668	540	328	373	213	850	1433	56	33	99	13.3
8th	11.08.20 – 17.08.20	799	855	600	443	468	345	644	1110	17	10	314	15
9th	18.08.20 – 24.08.20	829	527	466	276	434	116	741	818	15	3.33	30	12
10th	25.08.20 – 31.08.20	999	738	507	317	546	218	829	1490	96	23.3	144	5
11th	01.09.20 – 07.09.20	923	998	614	536	662	268	801	1858	8.5	0	209	14

Week No.	Date	33(A)		33(B)		33(C)		33(D)		33(E)		33(F)	
		mV	*μA*	*mV*	*μA*	*mV*	*μA*	*mV*	*μA*	*mV*	*μA*	*mV*	*μA*
12th	08.09.20 – 14.09.20	859	973	618	603	376	312	943	2182	17.3	0	87.1	8.3
13th	15.09.20 – 18.09.20	731	1160	611	593	597	335	920	1510	17	2.5	27	2.5

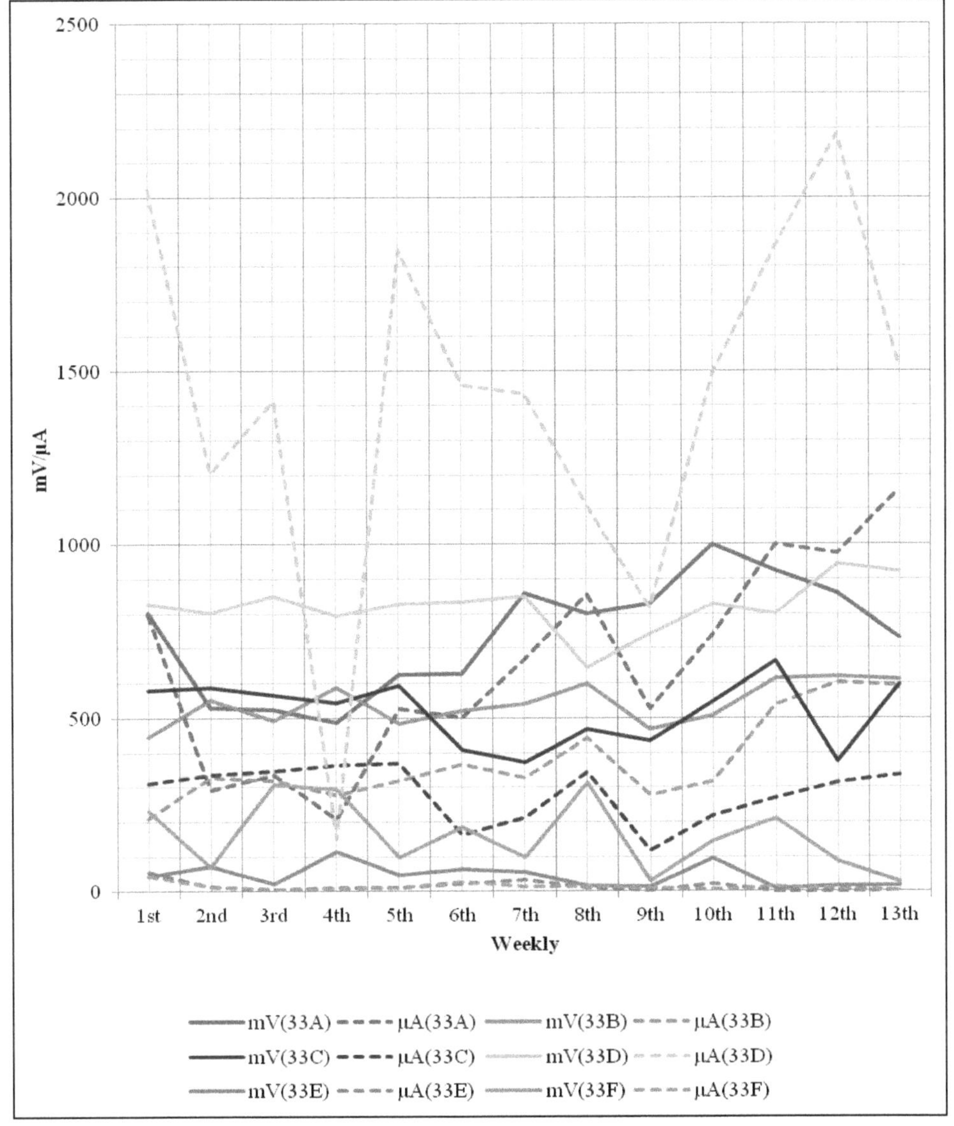

Figure 38: Pattern of Galvanic Current Flow through Roots of Papaya Seedlings under different Electrode Combinations.

Table 38: Measurements of Growth of Plants of Experiment 33

Date		22.06.20 (First Day)	24.07.20	19.08.20	24.08.20	31.08.20	07.09.20	15.09.20	23.09.20
33(A)	Length (mm)	211	216	330	350	380	400	450	460
	Width (mm)	-	3	8	8	10	13	13	13
33(B)	Length (mm)	158	240	540	540	710	750	850	930
	Width (mm)	-	6	14	16	20	22	23	26
33(C)	Length (mm)	153	210	440	470	565	600	640	700
	Width (mm)	-	5	12	17	19	21	21	21
33(D)	Length (mm)	201	270	560	600	690	760	840	940
	Width (mm)	-	6	18	20	22	26	28	30
33(E)	Length (mm)	200	207	214	220	225	230	270	300
	Width (mm)	-	4	7	7	7.5	9	9	9
33(F)	Length (mm)	210	260	550	590	640	640	680	720
	Width (mm)	-	6	15	18	21	21	21	21
33 (Control)	Length (mm)	72	74	75	77	85	100	135	160
	Width (mm)	-	5	5	5	5.5	6	6	6

The growth of seedlings (both in length and girth at the base) was recorded as follows (Table 38, Figure 39).

**Figure 39: Differential Growth in Papaya Seedlings from Left to Right –
(i) Untreated seedlings (Left), (ii) Galvanic current treated
seedling (Middle and Right).**

The seedling of ground soil died on 10.07.20 due to breakage of stem.

The highest growth during the period of 24.07.20 to 23.09.20 has been observed in 33 (D) (940 mm/30 mm) followed by 33 (B) (930 mm/26 mm); 33 (C) and 33 (F) (700 mm/21 mm and 720 mm/21 mm respectively), 33 (A) (460 mm/13mm); 33 (E) (300 mm/9 mm) and lowest in control (160 mm/6 mm).

Though these difference in growth of papaya plants could not be attributed to the intensity of galvanic current under which these plants were exposed and the leaves of 33 (C) and 33 (F) turned yellowish and curling due to limited space for growth of root system, yet as regards metals of electrode combinations, Cu and GI in 33 (D) was the best followed by Zn and Al in 33 (B) as the electrodes spacing between the plants were almost the same, though current intensities between them were different. Further investigations on this line are required to be carried out.

Comparative Evaluation

Comparative evaluation with the Prior Arts with that of the New Technique for Enhancing Germination Rate and Plant Growth with Galvanic Current Application

Patent Document

1.   Publication number: CN201510741107A

Weak Current Germination Accelerating Method for Ginkgo Tree Species

The applicant (Hexian zhenglin, Nursery stock specialize) claims a weak current could accelerate germination method of silver apricot tree, relating to the nursery stock breeding technology area. For accelerating germination in box is separately connected with power source and a current intensity of 15A was used for 25 seconds at an intervals of 15 minute for 4 – 6 times.

Comparison with our Experiments

In the present invention, no external power source was used. Galvanic current was generated around plant roots by placing suitable electrodes around the plant's root, which develops electrochemical cells (galvanic cells). Galvanic cells derive its energy by spontaneous redox reactions. Galvanic cells convert chemical potential energy into electrical energy by oxidation - reduction reaction, which was self induced by suitable electrodes around plant roots. The present invention is no way comparable to the invention published under publication no. CN 201510741107A, which described in the text that selected seeds were put in insulated sprouting box, set symmetrically on two sides of electrode plate. Seeds were sprayed with germination acceleration liquid every 2.5 hours. Two electrode plates were installed in the box connected with anode and cathode of the power source. The current intensity applied to the seed reported to be micro – current 15 mV A electrifying for 25 seconds for each time and after every 15 minutes once after each spraying of germination fluid for 4 – 6 times. After 24 hours seeds were taken out of sprouting chamber and kept in 25° - 30°C and 80 per cent relative humidity continuously for 4 days after sprouting. No follow up was reported to have been made after sprouting.

In our invention galvanic current coverage was given to germinating seeds with in soil or other moist media only with water, just as farmers do in the field in a conventional way. In our invention follow up with or without galvanic current coverage have been made for the germinating seeds to grow into plants, bear flowers and fruits and the distinct differences have been identified.

2.   Publication Number: KR2013118359A:

The Plant growth improvement method using electrostimulation

The inventor, Kim J. H claimed that his invention relates to a method for improving the growth of plant using electrical stimulation. The seeds of plant were dipped in a water tub for 3 – 18 hours at 17° - 27°C.

Electro-stimulation of 12 V, 80A for 20 seconds was applied to the seeds absorbing moisture with the help of two electrodes maintaining a gap of 40 – 60 mm. It was claimed that during electro – stimulation to the water absorbed seeds leads in generation of ultra sonic waves in which seeds were processed for 60 to 180 minutes. It follows a further thermal stimulation phase in which seeds in a separate tub are required to be put into a constant temperature of 25° - 45°C for 50 – 60 minutes.

Comparison with our Experiments

The present invention is not so complicated as of Kim J. H.

Seeds were exposed to galvanic current without any pre- and post- treatment as indicated in prior arts. As in a conventional way seeds are sown in soil or other moist media and were exposed to self induced galvanic current produced by suitable electrodes from the very beginning till germination and further growth of seedlings without any application of outside power source. Hence the present invention is entirely different from than of invention under the patent no. KR2013118359A

3.   Publication Number: KR2015187170A

Plant Growth Promotion Apparatus

Lim Jong and Kim Myong Su claims that their invention relates to a plant growth promotion apparatus which sterilizes fungi and bacteria by inserting an electrode in a pot in which plant was growing and applying electric current to soil accommodated in the pot, transmits electric stimulation to root of the plant through the soil, helps promotion of plant growth, and also has a beautiful external appearance at the same time.

Comparison with our Experiments

In the invention under patent No. KR2015187170A, electrical stimulation from outside, *i.e.* electric power received from collimator (possibly photoelectric cell where light received was converted to electric power) was used to apply in the soil with the help of electrodes, in a flower pot according to the predetermined program, which helped in promoting growth of plant by sterilizing fungi and bacteria.

No mention has been made about the intensity of such electric stimulation provided in the soil.

The invention is entirely different from the present invention applied for patent. In the present invention, the self induced galvanic current around plant roots was utilized for stimulation of germinating seeds, and enhanced growth of seedlings, early flowering and fruiting in fruit bearing plants and flowering plants.

4.   Publication Number: JP2008092903

Methods of Pre-Seeding Treatment for Plant

National Agriculture and Food Research Organization in their invention applied electric potential between a pair of electrodes arranged with a prescribed interval to form an electric field between electrodes. The plant seeds to be germinated and grown are retained within the electric field between the electrodes for a predetermined time prior to seeding, then are sowed. A DC or AC power source was used. In case of DC, the voltage applied between the electrodes was the optimum electrostatic field strength, set in the range of 1.3 KV to 15 KV depending on the type and amount of seeds.

In AC they used a low frequency of 500 Hz or less and a low voltage of 6 KV or less. Holding time in electric field was 30 – 60 seconds or lower. A deep cooling treatment at a low temperature of -10°C to -30°C was reported to be necessary.

Comparison with our Experiments

The present invention for which patent application was made was entirely different from the invention under patent no. JP2008092903 due to;

(a) In the present invention only self-generated galvanic current was used for a prolonged period, which were mostly in the range of millivolts and microamperes.

(b) No treatment either hot or cold was applied to the germinating seeds. Seeds were treated in ambient temperature as done in the conventional way.

5. Publication Number: DE2841933A1

Electrical Stimulation for Cell Growth Promotion - By applying direct current in pulses or with continuous polarity changes, to root area or to media contg. Nutrients

KABEL METALLWER in his invention claims that the growth of vegetables, animal and bacterial cells and tissues in beneficially affected by applying to the root area or to media containing nutrients, a direct current. According to him DC voltage is preferable. Pulsating DC with pulses of steep edges was also used.

Magnetic field generated by the flowing DC serves to influence the root areas or the media containing nutrients. He advocated that one electrode can be in the shape of a plate cover the bottom of the plant receptacle, the other is a rod inserted near the top. A still greater stimulation of growth can be achieved if the soil near the roots is intermingled with nutrients over 1 mu size in finely dispersed form.

Comparison with our Experiments

The inventor of DE 2841933A1 has not mentioned the DC and pulsating DC voltages which was applied to the plant. From the arrangements of electrodes mentioned in the text it appears that the plants were treated individually (the rod electrode inserted on the top). This is not clear.

So the invention under the patent no DE2841933A1 is not at all comparable with the present invention for which patent is being applied due to reasons mentioned above.

6. Publication Number: US20130318866

Apparatus and Method for Biological Growth Enhancement

Inventors of the patent claimed that their apparatus and method will benefit for biological growth enhancement, which includes seeds, fungus, bacteria and the like. High voltage (3000 – 7000 V) DC with low current were used for 5 – 10 minutes of 3 KV, 6 KV and 11 KV. In case of wheat seeds (200 nos. in a petridish) were hydro – primed, exposed to high voltage electric field and prepared for germination. The resulting sprouts were larger than those of inserted ones.

Comparison with our Experiments

In the present invention, no external high or low voltage was used. Only self generated galvanic current around germinating seeds, seedlings and plants were used under ambient conditions. Therefore the present invention is quite different from that of the invention under patent no. US 20130318866.

Non-Patent Document

1. M.B. Patil

Effect of Electrocultre on Seed Germination and Growth of *Raphanus sativus* (L) Mukundraj B. Patil

The author reported that seeds of *Raphanus sativus* (L) were supplied 3V, 6V and 9V of electricity for 10 mins daily. 95 per cent germination with 9V, followed by 90 per cent germination with 6V and 80 per cent germination in control and the pot supplied with 3V was obtain and the author considers supply of 9V electricity was most effective to affect the growth parameters.

Comparison with our Experiments

The author in abstract has not mentioned how the electricity was supplied to the plant to affect germination and for how many days.

Since, in the present invention, no electricity was supplied externally it is not comparable with the findings of the author (M. B. Patil).

2. F. Molamofrada *et al.*

The Effect of Electric Field on Seed Germination and Growth Parameters of Onion Seeds (*Allium cepa*)

Authors exposed onion seed samples (*Allium cepa*) to electric field of 2, 4, 9 and 14 kV/cm for 15, 45, 80 and 150 seconds. Treated seeds were transferred to the lab where growth characteristics were studied. The maximum increase in germination percentage and seedling height was in 9 kV/cm field intensity. Maximum increase in germination rate and dry weight of seedlings was observed in field intensity 2 kV/cm. 15 – 45 seconds exposure reported to be the optimum for maximum germination rate.

Comparison with our Experiments

The experimental procedure reported in the abstract is not clear about the treatment of onion seeds, the method adopted, comparison with non treated ones and the media in which the treated seeds were grown.

In fact, in warm, humid temperature onion seeds sprout in ambient condition. The effect of electric exposure could not be established specifically.

However, the non – patent document of F. Molamofrada is not at all comparable with the present invention, where no external electric current was applied. Only self induced galvanic current was used.

Comparing the prior arts and non-patent documents with that of the new technique for enhancing germination rate and plant growth with galvanic current application mentioned in the foregoing paras it is established that the present invention for which patent application was made entirely different due to the following reasons.

1. In the present invention, no external power source was used to stimulate the germinating seeds. No pre – or post hot or cold treatment was required to accelerate the germination rate and growth of plant.

   Self induced galvanic current was used either continuously or for a specific period to obtain higher rate of germination, reducing germination time and growth of seedlings and plants.

2. Placing suitable electrodes around germinating seeds and roots of seedlings and plants electrochemical cells were created to produce a voltage making electrons from a spontaneous reduction oxidation reaction.

3. The galvanic current stimulation around soaked seeds might have produced ionic changes in the growing seed tissue, as happened in medical galvanism to promote tissue healing in human.

4. The galvanic current generated around plant roots help to localize adequate concentrations of essential plant nutrients in plant available form. It accelerates the transport of mineral ions by ion exchange and also by stimulating root hair cells for larger surface area for quick absorption process.

5. The electrochemical potential that develops around the root, the signal of which is sensed near the interior of the root further stimulate the growth of plant.

6. For this advantage suitable types, size and shape of electrodes have to be chosen in order to obtain optimum current intensities in the range of 480 – 920 mV and 180 – 1000 µA in a sustainable manner.

7. The invention has a scope for increasing the rate of germination reducing the incubation period and vigorous growth of plants bearing large number of healthy fruits.

8. The technique was applied in 20 species of plant seeds, seedlings plants for nearly four years and the positive response was attained 70 – 80 per cent cases depending on species and electrode configuration.

This chapter is about how our ongoing research (Effect of Galvanic Current on Plant Growth) differs from the patented works which are very similar to this one. Moreover, the findings of all the above mentioned researchers indicate that galvanic current do have some beneficial effects of plant germination and growth.

Epilogue

The utility of the new technique for enhancing germination rate of seeds and plant growth with galvanic current application was assessed by three ways as detailed below;

(a) By observing the rate of germination of different varieties of plant seeds;

(b) Survival and growth rate of seedlings thus produced, either with galvanic current coverage, or without coverage.

(c) The growth of seedlings and plants were determined by the length of plants above ground level (media), girth of the plant at the base of the stem (5 cm above ground level) and number of main branches (not the sub – branches).

Studies have been also made about the production of sub – soil galvanic current by inserting electrodes of different metals of different shapes and dimensions, distances between the electrodes and from the plant's root system. The production of galvanic currents between the electrodes (mV and µA) were measured daily after 24 hours.

In all the experiments controls were in identical conditions with that of experimental ones except without electrodes.

The role of different media (water, moist absorbent, soil) for production of galvanic current and its influence on germination and plant growth in primary stages were also studied.

In Bengal gram (*Cicer arietinum*), 100 per cent germination with 7 – 35 branches was observed under the influence of galvanic current (470 – 560 mV;

208 – 1000 µA) produced by aluminium electrodes (6 X 12 cm²) in a water media as against 70 per cent germination with 6 – 26 branches in control in 21 days.

With zinc and aluminium electrodes in soil media in electrically insulated plastic trays producing a potential difference of 216 – 533 mV and current intensity of 38 – 200 µA could germinate 70 per cent sunflower (*Helianthus annuus*), 66 per cent cucumber (*Cucumis sativus*) and 40 per cent snake bean (*Vigna unguiculata*) as against 30 per cent sunflower, 10 per cent cucumber and 10 per cent snake bean in control within 2 days. On the 7th day sunflower grew to a length of 10.8 – 12.4 cm; cucmber to a length of 9.7 – 11.8 cm and snake bean 9.5 – 15.2 cm as against 1 -12 cm in sunflower, 5.8 – 6.8 cm in cucumber and no growth (died) in snake bean in control.

A positive influence of galvanic current produced by zinc electrodes (Zn + Zn) (662 mV; 310 µA) have been observed with respect to germination of sunflower (80 per cent), paddy (*Oryza sativa*) (50 per cent) as against aluminium electrodes (Al + Al) (657 mV; 300 µA) whereas no germination of sunflower was observed and only 40 per cent paddy germinated in control.

Applying galvanic current through water and soil with the help of aluminium and copper electrodes; 60 per cent corn (*Zea mays*) germinated under the influence of 542 mV and 480 µA and grew to 12 – 33.4 cm (1.54 cm/day). But when aluminium and galvanized iron electrodes were used the galvanic current intensities were 190 mV and 470 µA which effected 40 per cent germination of corn with an increment in growth of 1.37 cm/day. 25 per cent corn germinated when copper and galvanized iron were used as electrodes producing 900 – 996 mV and 680 – 1780 µA. The increment in growth under this combination was more (1.86 cm/day), possibly due to lesser seedling density. The germination of corn seed was only 20 per cent (lowest) in control (without galvanic current exposure) and the seedlings did not survive.

After 12 days in soil media with galvanic current coverage (as mentioned above) corn seedlings were transferred to the ground soil removing the galvanic current so as to allow to grow like normal plants. After 63 days of transferring corn seedlings to ground soil the plant grew to a length of 66 – 127 cm bearing two corns in each plant which has been exposed to galvanic current with Al + Cu electrodes. Al + GI treated plants bore single corn in each plant (71.1 cm). Highest length of 142.2 cm was attained in plant, treated with Cu + GI electrodes, with 4 branches bearing 4 corns (1 corn in each branch) at a time. It is thought that with higher intensity of galvanic current (900 – 996 mV and 680 – 1780 µA) generated by copper and galvanic iron electrodes an exceptional genetic modification might have taken place (multiple branching in monocot stem and bearing fruits in multiple branches) which need further investigation for beneficial cultivation.

100 per cent germination and survival of tomato (*Solanum lycopersicum*) was effected when the seeds were put under the influence of galvanic current through the moist (soaked with water) cotton with the help of bent copper and aluminium plate and two bent aluminium plates in two plastic bowls along with control without electrodes. Within 3 days the seedlings attained 0.6 – 8 cm in the first bowl; 2.7 – 5.5 cm in the second bowl with aluminium electrodes. The galvanic current intensities in the first bowl with Cu + Al electrodes were 524 mV – 594 mV and 490 – 780 μA, while in the bowl with two aluminium electrodes current intensities registered 115 – 130 mV and 0 μA. In the control of the same set of experiment 75 per cent seed germinated and the seedlings were of 4.1 – 4.8 cm.

After 9 days tomato seedling were shifted into ground soil with copper and aluminium electrodes around the seedlings in the first set and two aluminum electrodes around the second set.

Galvanic potential around the tomato seedlings in ground soil fluctuated between 553 – 760 mV and 120 – 470 μA in the first set with Cu and Al electrodes; whereas in the 2nd set with aluminium electrodes, the variation was between 5.2 – 138 mV and 1.43 – 45.94 μA.

The survival of tomato seedlings after 148 days with Cu + Al electrodes was 75 per cent while in Al + Al electrodes was 50 per cent and in control 30 per cent. Ripe tomatoes were harvested after 150 days of germination.

Tomato plants in the first set (Cu + Al) attained a length of 193 – 195 cm, with 9 – 13 branches in each plant and bore 26 ripe fruits weighing 6.02 – 58.76 gm each with a total weight of 751 gm. The dry weight of tomato plants (excluding fruits) ranged from 280 – 325 gm for each plant.

In the second set with two aluminium electrodes only, plants attained 180 – 190 cm, with 3 – 7 branches in each plant. Plants did not bear any fruits (tomatoes) and the weight of biomass (foilage) ranged between 100 – 350 gm each.

Plants in control attained a length of 170 – 240 cm with 6 – 7 branches in each. 16 ripe tomatoes weighing between 1.46 – 30.95 gm each and total weight of 222.4 gm was harvested. Weight of biomass (foliage) was 250 – 350 gm each.

As such, the galvanic current exposure (553 – 760 mV; 120 – 470 μA) from Cu and Al was found to have a beneficial effect on growth and fruit bearing capability of tomato plant.

In order to find out the optimum electrode combinations for growth of papaya seedlings (*Carica papaya*), seedlings between 15 – 16 cm length have been planted in ground soil with Zn + Al, Cu + Al, Cu + GI, Cu +Cu and Al + Al electrode combinations. The following table will indicate the pattern of galvanic current flow around the seedlings root system and the growth of the plants (length and girth).

Electrode Combinations	Spacing between Electrode (cm)	Galvanic Current Flow Ranged		Initial Growth		Growth After					
						30 Days		60 Days		90 Days	
		mV	µA	Length (cm)	Girth (mm)	Length (cm)	Girth (mm)	Length (cm)	Girth (mm)	Length (cm)	Girth (mm)
Zn + Al	19.5	535 – 589	330 – 660	15.3	3.0	24	6	54	14	93	26
Cu + Al	19.5	573 – 618	310 – 330	15.5	3.5	21	5	44	12	70	21
Cu + GI	19.5	865 – 924	2160 – 2310	15.8	3.0	27	6	56	18	94	30
Cu + Cu	19.5	1.1 – 67	0 – 20	15.8	3.0	21	4	24	7	30	9
Al + Al	19.5	4.1 – 15	0 – 10	16	3.5	26	6	55	15	72	20
Control	-	-	-	15	3.0	16	4	17.5	5	21	7

It may be observed from the above table that Cu + GI and Zn + Al were the most optimum electrode combinations for enhancing growth of papaya plant followed by Cu + Al and Al + Al.

The aforementioned informations were results of some of our experimental studies. Studies have been made on 20 species of plant seeds and seedlings and not all the species respond to the galvanic current in the same way. Each species have its own way of adaptations and response to sub – soil galvanic current.

Electrode combinations and production of sub – soil galvanic current flow are the crucial factors for the positive response with regards to germination and growth besides the normal criteria of suitable temperature, humidity, space and light. Rectangular metal plates (18 gauge) of 6 – 12 cm width was found optimum of which at least 4 – 6 cm are required to be embedded in the soil around the seed/seedlings to be treated with galvanic current. The length of electrodes depend on the size of the germinating space (plots) to be covered under galvanic current exposure. In our trials we have used electrodes with length of 12 cm to 62 cm. The distance between the electrodes varied with the size and shape of the plots and the spacing of the seeds/seedlings in the plot to be treated in each case. In any case the distance between the seed/seedlings and the electrode should not exceed 30 cm in order to receive the optimum strength of current intensities to the root systems.

In fact experiments are required to be conducted on each of the plant species with different electrode combinations for optimum results. Our research unit has taken up trials at kitchen garden scale on different seasonal vegetable crops. Much work is required to be done in a systematic manner.

But the indications that have been obtained from 52 sets (115 experiments) over period of four and half years indicate that there are sufficient reasons to believe that this new technique galvanic stimulation may open a new horizon in agriculture practices in future, especially in a limiting farming space.

The findings of our experiments lead to conclude;

(a) The rate of germination is higher and quicker under galvanic current exposure.

(b) The growth of seedlings and plants are more under galvanic current stimulation.

(c) Each plant species require specific intensities of galvanic current to respond positively.

(d) Type, size and shape of electrode effect subsoil galvanic current production.

(e) Moderately moist clayey soil generates galvanic current steadily.

(f) Dissimilar metals as electrodes produce higher and steady current flow.

Our Ongoing Experimental Work to Find Out Influence of Galvanic Current on the Growth of Cauliflower (*Brassica oleracea var. botrytis*) and Cabbage (*Brassica oleracea var. capitata*); Experiment (Left one) and Control (Right one).

Ongoing Work of Multi Species Farming under Influence of Galvanic Current.

References

Benoliel, R.,Gary MHeir, &Eli Eliav(2008), Neuropathic Orofacial Pain.In: Yair Sharav& Rafael Benoliel, *Orofacial Pain and Headache*, (pp. 255-294).Elsevier. https://doi.org/10.1016/B978-0-7234-3412-2.10011-2

Bibikova, T., Gilroy, S. Root Hair Development. *J Plant Growth Regul* 21, 383-415 (2002). https://doi.org/10.1007/s00344-003-0007-x

Ciechanski, P., &Adam Kirton (2016), Transcranial Direct-Current Stimulation (tDCS):Principles and Emerging Applications in Children. In: A. Kirton, Donald L. Gilbert (eds), *Pediatric Brain Stimulation*(pp. 85-115).ELSEVIER. https://doi.org/10.1016/B978-0-12-802001-2.00005-9

Iwabuchi, A., Yano, M. & Shimizu, H. Development of extracellular electric pattern around *Lepidium* roots: its possible role in growth and gravitropism. Protoplasma 148, 94-100 (1989). https://doi.org/10.1007/BF02079327

Milady, Joel Gerson & Janet Dangelo (2008) Facial Machines. In: *Miladys Standard Esthetics Fundametals* (10th ed).

Press, Joel M., & Deborah A. Bergfeld, (2007),Physical Modalities. In:Walter R. Frontera, Stanley A. Herring &Lyle J. Micheli, et. al., *Clinical Sports Medicine: Medical Management and Rehabilitation*(pp. 207-226).Elsevier. https://doi.org/10.1016/B978-141602443-9.50019-2

Sakaguchi, Ronald L., & John M. Powers, (2012), Fundamentals of Materials Science. In:*Craig's Restorative Dental Materials*(13th ed., pp. 33-81). Eslevier.https://doi.org/10.1016/B978-0-323-08108-5.10004-0

Shu Ezaki, Kiyoshi Toko, Kaoru Yamafuji, & Fujio Irie (1988), Electrical Potential Patterns around a Root of the Higher Plant, IEICE TRANSACTIONS(1976-1990),Vol.E71, No.10, (pp. 965-967).

Takamura T. (2006), Electrochemical Potential Around the Plant Root in Relation to Metabolism and Growth Acceleration, In: Volkov A. G. (eds) *Plant Electrophysiology*. Springer, Berlin, Heidelberg.

https://doi.org/10.1007/978-3-540-37843-3_15

Volkov A. G., Markin V. S. (2015), Active and Passive Electrical Signaling in plants. In: Lüttge U., Beyschlag W. (eds) *Progress in Botany*. Progress in Botany (Genetics – Physiology – Systematics – Ecology), vol 76. Springer, Cham. https://doi.org/10.1007/978-3-319-08807-5_6.

Page: 53

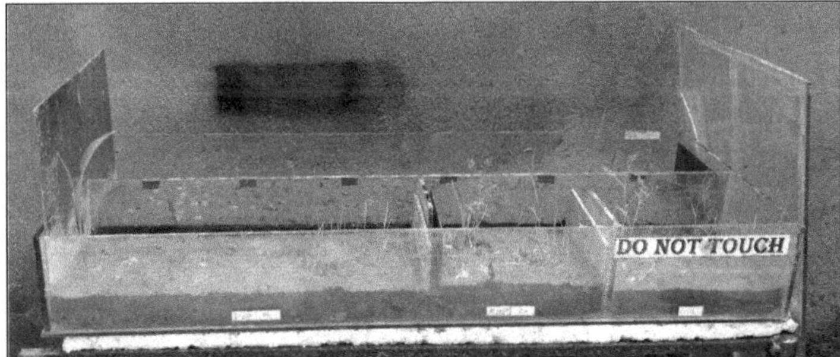

Page: 58

Page: 58

Page: 61

Page: 61

Page: 71

Page: 72   Page: 73

Page: 86

Page: 87

Page: 113

Page: 132

Page: 132

Index